Synoptic skills in
ADVANCED
CHEMISTRY

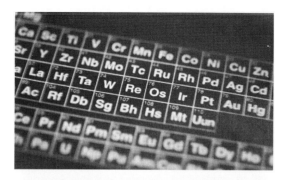

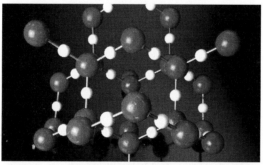

D0434255

GRAHAM HILL

Hodder & Stoughton

A MEMBER OF THE HODDER HEADLINE GROUP

ACKNOWLEDGEMENTS

I am grateful to the awarding bodies (AQA, Edexcel and OCR) for permission to reproduce a number of synoptic questions.

I should however point out that the responsibility for the suggested answers to these and other questions in the book is entirely mine.

Graham Hill

Orders: please contact Bookpoint Ltd, 130 Milton Park, Abingdon, Oxon OX14 4SB. Telephone: (44) 01235 827720. Fax: (44) 01235 400454. Lines are open from 9.00 – 6.00, Monday to Saturday, with a 24 hour message answering service. You can also order through website www.hodderheadline.co.uk

British Library Cataloguing in Publication Data
A catalogue record for this title is available from the British Library

ISBN 0 340 84457 4

First Published 2002
Impression number   10 9 8 7 6 5 4 3 2
Year   2008 2007 2006 2005 2004

Typeset by J&L Composition, Filey, North Yorkshire.
Printed in Great Britain for Hodder & Stoughton Educational, a division of Hodder Headline Ltd, 338 Euston Road, London NW1 3BH by JW Arrowsmith, Bristol.

PREFACE

New AS and A level Chemistry Courses were introduced in September 2000.

All of the revised specifications devote 20% of the total assessment towards synoptic skills, all of which is covered during the second year of the A level course. This means that 40% of the A2 examination is devoted to the assessment of synoptic skills.

The huge emphasis on synoptic skills in the A2 modules puts greater demands on students and teachers to develop and improve the relevant skills.

With this in mind, the aims of this book are:

★ to provide information, advice and guidance on synoptic skills and their assessment,

★ to help students develop and improve their skills in answering synoptic questions,

★ to help teachers appreciate the requirements of synoptic assessment.

I am grateful to my publishers, Hodder and Stoughton Educational and, in particular, Elisabeth Tribe (Director of Schools Publishing) for the invitation to write this book. I must also thank Helena Ingham (Editorial Assistant) and my wife, Elizabeth, for the helpful and efficient manner in which they have managed the whole project towards publication.

Graham Hill
July 2002

CONTENTS

CHAPTER ONE

GETTING THE MOST FROM THIS BOOK

Why you need this book

Synoptic skills and synoptic questions involve making connections and seeing links between facts and ideas in different areas of a subject. This is not so very different from the skills you use in everyday life. For example, when you plan a journey, you need to ensure that one stage connects smoothly with the next. By making connections between different areas of your chemistry course, you will begin to understand the subject much better and it will become more interesting.

New Advanced Subsidiary (AS level) and Advanced GCE (A level) Chemistry Courses were introduced in September 2000. The criteria for these courses state that A level specifications must devote at least 20% of the total A level assessment towards synoptic skills.

In fact, all of the five chemistry specifications [AQA, Edexcel, Edexcel (Nuffield), OCR and OCR (Salters)] devote exactly 20% of the total assessment to synoptic skills. What is more, all five specifications also delay all of their synoptic assessment until the A2 modules in the second year of the A level course. There is no synoptic assessment in AS Chemistry, normally studied in the first year of the full A level course.

As both AS and A2 each contribute 50% of the A level assessment and all of the synoptic component is carried out during A2, it means that 40% of the assessment of A2 units is devoted to synoptic skills – an exceptionally high percentage.

For some years now, A level examinations have included an element of synoptic assessment. During this time, teachers and examiners have become aware that students have answered synoptic questions less well than others. Given the emphasis in synoptic testing on making connections between different areas of the entire A level course, this lower performance is perhaps not surprising.

The increased emphasis on synoptic skills in the new specifications and their entire assessment in the A2 modules does, however, put greater demands on students and teachers to develop and improve the relevant skills.

The aims of this book are therefore:

★ to provide information, advice and guidance on synoptic skills and their assessment;

★ to help students develop and improve their skills in answering synoptic questions;

★ to help teachers appreciate the requirements of synoptic assessment.

How to use this book

The following suggestions will help you to get the most from this book.

1 Follow the advice on study skills and revision techniques in the next part of this chapter. Good study skills are the key to academic success.

2 Make sure you appreciate:

★ what is meant by synoptic skills and how your awarding body (i.e. AQA, Edexcel or OCR) will incorporate synopsis into the specification for A level Chemistry (chapter 2).

★ when (at what stage of the course), where (which examination paper or coursework) and how (what kind of test/assessment and what sort of questions) synoptic skills will be assessed in your course (chapter 3).

3 Improve and extend your synoptic skills by:

★ using the strategies suggested in chapter 4; and

★ analysing the various types of synoptic questions and appreciating how they should be answered (chapter 5).

4 Finally, practise answering synoptic questions using the specimen examination units/assessment units in chapter 6 and then check your own answers against the suggested answers and mark schemes in chapter 7.

Study skills and revision techniques

Good study skills and effective revision techniques are essential to academic success. Without the right approach and commitment in this area, you are unlikely to reach your potential in any test or exam. Here, then, are some important suggestions on how to study, followed by advice on revision.

(i) Organising your time

Most schools and colleges will provide between four and five hours of teaching per week for each A level subject. In addition to this, they will expect you to spend at least the same amount of time in private study, whether on site or at home. Without this additional commitment, your progress and achievements will not match your potential.

With this in mind, it is well worth reviewing your weekly commitments from the beginning of your A level studies. You will need to consider and plan realistic allocations of time to private study, exercise, relaxation and sleep – in addition to your studies at school or college.

You may want to take on a part-time job. If this is so, remember two things:

★ it would be foolish to allow the short-term gains of a few extra pounds in your pocket *now* to jeopardise the long-term benefits of greater success at AS and A level;

★ a report for the *Times Educational Supplement* suggested that the academic studies of A level students began to suffer if they were employed in a part-time job for more than 7 hours per week.

The efficient organisation of your time should also involve the effective use of spare moments. For example, you could be making notes, looking through your own (home-made) revision cards or reading round the subject when travelling to school or college or waiting for an appointment. These short sessions of work will add up to many hours of useful study during the whole of the A level course.

(ii) Organising your studies

Teaching time will, of course, be organised and planned by your school or college, but the rest of your studies – checking your notes, homework, additional note making, revision, etc. – are your responsibility.

The quality and presentation of your notes is very important.

★ Make clear and concise notes.

★ Make headings and sub-headings prominent.

★ Highlight key points and definitions in some way.

★ Draw tables and diagrams and make lists to help your understanding and recall of key facts and information.

★ Make an overall plan and file your notes accordingly – ideally in order of the units (modules) and sub-units in the A level specification.

After lessons, it is a good idea to check your notes using your textbook(s) and to fill in any gaps. Never underestimate the importance of note-making, summarising and checking. These activities of 'pen on paper' are powerful aids to your assimilation and understanding of the chemistry course.

Most teachers and lecturers will set work for your private study and homework. It should go without saying that this work should be carried out fully and seriously. It may involve reading, note-making, planning, preparation, answering past examination questions or revision.

For the most part, this set work will be an essential and planned element of the AS and A2 programmes, geared towards the best possible conditions for your success.

If the set work involves reading a textbook, do remember that this is not like reading a magazine or a novel. Emphasis must be on understanding the text, sentence by sentence. So, some parts of the book, such as equations and theoretical ideas, will require careful study and slow checking.

Finally in this section, remember that your teacher is not the only readily available source of ideas, help, information and discussion about chemistry and your chemistry studies. Discuss homework problems, approaches to practical work, applications of chemistry and your understanding of difficult ideas with other students. Just expressing your difficulties and articulating your ideas will help your understanding and increase your interest.

Hints for revision

The most important aim of AS and A level Chemistry examinations is to assess your knowledge, your understanding and your skills – what you know, what you understand and what you can do.

The key to all this, to your revision and to ultimate success, is to ensure that you have understood what you have studied. If you understand a topic, it is much easier to revise and remember the key facts.

In order to fully understand a topic, you must do more than just read your notes or read your textbook(s). Your studies and revision must be *active* rather than passive. Here are some approaches and activities which will help to keep your learning active and interesting.

★ Underline or highlight important words, statements and definitions.

★ Make lists of key words or key facts.

★ Write out and learn important definitions. Writing notes will help you to remember key points.

★ Draw diagrams to summarise important topics. Label the diagrams and write notes at the side. Diagrams are a powerful way to reinforce your memory and your understanding. Most people find it easier to remember things from pictures than from words alone.

★ Summarise important ideas, explanations and processes. Sometimes it is helpful to make a flowchart to summarise a sequence of ideas or events. The flowchart is simply a series of words or short sentences connected by arrows.

★ Keep your lists, notes, diagrams and summaries. All these items can be looked at, reviewed and annotated again and again. Diagrams and summaries that *you* have made will jog your memory and understanding very quickly. If your notes are disorganised, you will gain little from them. But with concise notes, clear summaries and good diagrams, you will increase your long-term knowledge and understanding significantly.

★ Try to copy your summaries and diagrams from memory. This won't be easy at first. You will need to look at your original to refresh your memory, but don't be discouraged. Testing yourself regularly on work previously revised is an important part of exam preparation.

★ Answer past exam questions and check the answers. Once you have a reasonable knowledge and understanding of the topics concerned, answering past exam questions is one of the best ways to revise.

★ If you don't understand something ask your teacher. Make a note of these difficulties or queries, otherwise you will not remember them. Your teacher wants you to succeed as much as you do. He or she will be delighted to help you with your revision, particularly if your motivation and commitment are clear.

Keeping fit and well during revision

Keeping fit and well during revision and the examination period is just as important as keeping fit and well during training for a sports event.

★ Don't overdo your revision. Allocate a realistic period for your revision schedule and a reasonable time for revision each day. Remember that preparing for an exam is like a marathon – not a sprint!

★ In any study leave from school or college, set aside a regular time for studying each day and stick to it.

★ During revision sessions, study for 40 to 50 minutes, then take a break for 10 minutes or so. Continue revising for another 40 to 50 minutes, followed by another short break.

★ During your breaks, do something different to take your mind off revision.

★ While you are revising, avoid distractions from friends, family, radio and television.

★ Make sure that you have regular meals and that you get enough exercise, relaxation and sleep.

★ Don't try to do much revision the evening before an exam. Just read through the notes and summaries to reassure yourself.

★ Finally, keep a sense of balance and remember that AS and A level examinations are not the end of the world!

The examination, assessment units and unit tests

The different awarding bodies use different names for the various examination components (examination papers). Edexcel refer to the examination components/papers as 'unit tests', whilst AQA and OCR refer to them as 'assessment units'.

Terms used in the unit tests/assessment units

A number of terms, commonly used in unit tests, are listed below. It is important that you understand clearly the meaning of each of these terms so as to answer questions appropriately.

★ Calculate – carry out a calculation, show your working and give the answer to the correct number of significant figures with appropriate units.

- ★ <u>Compare</u> – point out similarities *and* differences.

- ★ <u>Define</u> – give a precise statement of what is meant by a particular term.

- ★ <u>Describe</u> – provide a statement covering the main points or observations. An explanation is *not* necessary.

- ★ <u>Discuss</u> – describe and point out the key points/different opinions on an issue.

- ★ <u>Distinguish between</u> – point out differences only.

- ★ <u>Explain</u> – give reasons, with reference to the chemistry involved. A description is *not* required.

- ★ <u>Give</u> – give a concise, factual answer.

- ★ <u>Name</u> – give a concise, factual answer.

- ★ <u>Outline</u> – give a brief account.

- ★ <u>Predict</u> – use your chemical understanding to give an appropriate answer.

- ★ <u>State</u> – give a concise, factual answer.

- ★ <u>Suggest</u> – use your chemical knowledge to put forward an appropriate answer, possibly to an unfamiliar situation.

- ★ <u>What/Why/How</u> – direct questions requiring concise answers.

Whatever the terms used in exam questions, you must read each question *carefully*, think about your response and remember that the marks allocated to a question are a guide to the depth of answer required.

Types of examination questions

Essentially, there are four types of questions that are used in AS and A level Chemistry examinations:

- ★ short-answer questions;
- ★ structured questions;
- ★ extended prose/directed essays;
- ★ objective questions.

It is important that you should appreciate these different types of question.

Short-answer questions

Short-answer questions are not very common at AS and A level, although a number of them can be linked together in structured questions. There are several types of short-answer questions. The required answer may be a single word, several words or a sentence.

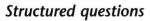

Structured questions

Structured questions are the most common questions on AS and A level papers. They are by far the most prominent type of question for all five chemistry specifications.

A structured question usually consists of some introductory information, followed by a series of questions based on that information. Sometimes, the information concerns social, environmental, economic or industrial aspects of chemistry. It is essential that you read the introductory information carefully before starting to answer the questions.

In structured questions, spaces are usually left for the answers on the question paper itself. Very often, the marks for each part of the question are also shown. The space left and the marks shown are a guide to the length of the answer required and the time you should spend on each part. Don't feel you must fill the space left for the answer, but if your answer is much too short, think again. On the other hand, if your writing is average-sized and you have insufficient space for the answer, your answer is probably more detailed than necessary.

Typically, the parts to structured questions involve:

★ naming parts on or completing diagrams;
★ writing balanced equations;
★ completing tables;
★ predicting reactions;
★ plotting graphs;
★ interpreting data;
★ performing calculations.

Extended prose/directed essays

Extended prose questions can be used on their own, or as part of longer structured questions. In these questions, you are expected to write four, five or six sentences.

All the specifications include extended prose questions and within these questions you should present information, descriptions and arguments clearly and logically, taking care over your grammar, punctuation and spelling.

The AQA, Edexcel (Nuffield) and OCR specifications also include much longer extended prose questions, carrying 10 or more marks. These questions are probably best described as directed essays as they indicate pretty clearly what is required.

Bear in mind the following points as you answer directed essays.

★ Make your points clearly and concisely, illustrating with examples if it is appropriate.
★ Avoid repetition and don't begin your answer by repeating the question.
★ Keep your answer relevant.
★ Cover all the topics addressed in the question.

Objective questions

Only one specification – that from AQA – uses objective questions as part of its assessment. Objective questions require only a single letter for the answer. Two different types of objective question are used – multiple choice and multiple completion.

★ Multiple choice questions ask you to choose the one correct answer from four or five alternatives. For example:

Which **one** of the following substances can be oxidised?

A CaO

B NO

C SO_3

D SnO_2

Answer: **B**

★ Multiple completion questions provide four possible answers, one or more of which is/are correct. You must decide which of the possible answers is/are correct and mark **A**, **B**, **C** or **D** on the answer sheet as follows.

A if (i), (ii) and (iii) only are correct

B if (i) and (iii) only are correct

C if (ii) and (iv) only are correct.

D if (iv) alone is correct.

For example:

Ionic bonding is found in

(i) NH_4Cl

(ii) $NaAlH_4$

(iii) CH_3COONa

(iv) CCl_4

Answer: **A**

At first sight, these questions may seem easy because they involve very little writing. But don't be deceived! Many objective questions are carefully designed to test difficult ideas. You should prepare just as thoroughly for an assessment unit involving objective questions as you would for one involving longer questions.

Finally, never leave an objective question unanswered. If necessary, make an intelligent guess as you do *not* lose marks for a wrong answer. For example, in the question above you may know that CCl_4 does not contain ionic bonding. This means that the answer cannot be **C** or **D**. You would now have a 50:50 chance of guessing the right answer.

Preparing for the exam (unit tests)

Success in any exam often depends as much on good exam technique as on study skills throughout the course and revision prior to the test.

★ <u>Arrive for the exam in plenty of time</u>. This way you will avoid worrying about the possibility of missing the start.

★ <u>Come fully equipped</u> with pen, pencil, rubber, ruler, calculator and watch.

★ <u>Read the exam instructions and questions carefully</u>.

★ <u>Plan your time sensibly</u>. If you have to do four questions with equal marks in one hour, then you should not spend more than 15 minutes on any one question. If you get stuck on part of a question, leave it and come back to it at the end.

★ <u>Answer all the questions</u>. Even if you are unsure about your answer to a question, remember that it is usually very easy to gain the first one-third of the marks. Check that you have answered all the questions expected of you and that you have not missed any.

★ <u>Keep your answers relevant</u>. Answer the question you have been asked. Don't waffle, as this will only waste time and you will not have gained any marks.

★ <u>If you have to write at length</u>, say more than two or three sentences, then:
  – jot down in rough the key points;
  – decide the order of these points before starting to write;
  – use short, clear sentences.

★ <u>If a diagram is required</u>:
  – draw it clearly in pencil;
  – label it neatly in ink.

★ <u>If a graph is required</u>, make sure that you:
  – label the axes;
  – show scales on the axes;
  – include the units for quantities plotted along the axes;
  – show the points clearly;
  – draw the 'line or curve of best fit' smoothly.

★ <u>If a calculation is required</u>, make sure that you:
  – explain what you are doing, stating the principle or equation you are using;
  – show your working;
  – give the answer to the appropriate number of significant figures. If, for example, relative atomic masses required for a calculation are given to three significant figures (i.e. $A_r(Cl) = 35.5$), then your final answer should not be given to more than three significant figures.

- write the correct units with your numerical answer, e.g. the concentration of a solution will be 2.5 mol dm^{-3} or 2.5 M, not simply 2.5.

★ <u>Keep your writing neat and legible</u>. You will gain no marks if the examiner cannot read what you have written.

★ <u>If you make an error</u>, don't use Tippex. Cross out the error so that your original writing is still legible. Crossed-out work can sometimes work to a candidate's advantage.

★ <u>Check your spelling, punctuation and grammar.</u>

Finally, GOOD LUCK!

CHAPTER TWO

UNDERSTANDING SYNOPTIC ASSESSMENT

What are synoptic skills?

All five A level Chemistry specifications make the same statements about synoptic assessment. Essentially, synoptic assessment involves the drawing together of knowledge, understanding and skills learned in different parts of the A level course.

The emphasis of synoptic assessment is on understanding and application, rather than on recall. It will test your ability to draw together and integrate your understanding and skills with the knowledge which you have acquired throughout the whole course.

Here is precisely what the specifications say.

Synoptic assessment:

- **requires students to make connections between different areas of chemistry – for example, by applying knowledge and understanding of the principles and concepts of chemistry in planning experimental work and in the analysis and evaluation of data.**

- **includes opportunities for students to use skills and ideas that permeate chemistry in contexts which may be new to them – for example, writing chemical equations, quantitative work, relating empirical data to knowledge and understanding.**

Let's consider each of these skills in turn, so as to appreciate exactly what synoptic testing will involve.

> **Students should be able to make connections between different areas of chemistry by applying knowledge and understanding of the principles and concepts of chemistry.**

This statement means that you should be able to use your knowledge of, say, the chemistry of iron and general principles of transition metal chemistry, electronic structure and bonding to answer a question such as the one below. This question is about (i) the formula, preparation, structure and bonding of the chlorides of iron, (ii) the electronic structures of iron and its ions and (iii) a simple test for chloride ions.

EXAMPLE A

A student prepared two chlorides of iron by carrying out two experiments in the laboratory.

a) In the first experiment, the student reacted iron with an excess of hydrogen chloride gas forming a chloride **A**, with a composition by mass, Fe: 44.0%; Cl: 56.0%.
(A_r: Fe, 55.8; Cl, 35.5.)

   (i) Identify compound **A**, including all of your working in your answer.
   (ii) Construct an equation for this reaction. *(3)*

b) In the second experiment, the student formed 8.12 g of a chloride **B** by reacting 2.79 g of iron with an excess of chlorine.
(A_r: Fe, 55.8; Cl, 35.5.)

   (i) Identify compound **B**, including all of your working in your answer.
   (ii) Construct an equation for this reaction. *(4)*

c) Write down the sub-shell electronic configurations of iron in

   (i) metallic iron;   (ii) compound **A**;   (iii) compound **B**. *(3)*

d) Aqueous solutions of **A** and **B** both contain $Cl^-(aq)$ ions. Describe a simple test for the presence of these ions. *(2)*

e) The chloride **A** has a much higher melting point (672°C) than that of the chloride **B** (220°C). An aqueous solution of chloride **A** is neutral whereas that of **B** is acidic. Explain what this information suggests about the structure and bonding in **A** and in **B**. *(4)*

(Total: 16)   OCR Specimen Question

Try this question for yourself and then check your answer with the specimen answer and mark scheme in chapter 7.

Non-synoptic questions will tend to focus on one particular topic (say, transition metal chemistry or structure and bonding or calculation of an empirical formula). Synoptic questions, however, require you to make connections between the facts, principles and concepts in different areas of the subject.

Let's now consider the second key skill associated with synoptic testing.

> **Students should have opportunities to use the skills and ideas that permeate chemistry in contexts which may be new to them.**

Again, you must be able to bring together different areas of chemistry, but the emphasis in this statement is on *skills*. The important skills which you will need to use in answering synoptic questions are:

★ planning experiments;

★ using, analysing and interpreting data in terms of chemical principles and concepts;

★ predicting properties, reactions and changes;

★ appreciating the social, environmental, economic and technological applications and implications of chemistry.

These requirements are well illustrated by the following question which brings together an understanding of (i) titration curves, (ii) acid dissociation, (iii) empirical, molecular and structural formulas, (iv) chiral compounds and (v) mass spectrometry, whilst also testing your skills in planning, analysing, interpreting and predicting.

EXAMPLE B

A carboxylic acid, **A**, contains 40.0% carbon, 6.70% hydrogen and 53.3% oxygen by mass. When 10.0cm^3 of an aqueous solution of **A** containing 7.20 g dm^{-3} was titrated against 0.050 mol dm^{-3} sodium hydroxide, the following pH readings were obtained.

Volume of NaOH /cm^3	0.0	2.5	5.0	7.5	10.0	14.0	15.0	16.0	17.5	20.0	22.5
pH	2.5	3.2	3.5	3.8	4.1	4.7	5.2	9.1	11.5	11.8	12.0

a) (i) Plot a graph of pH (on the y axis) against volume of NaOH.
Use the graph to determine the end-point of the titration. Hence, calculate the relative molecular mass of **A**. (8)
(ii) Calculate the value of K_a for **A**. (3)

b) (i) Use the data at the start of the question and your result from part **(a)**(i) to calculate the molecular formula of **A**. (2)
(ii) Suggest two structures for **A** that are consistent with the information above. (2)

c) (i) Describe **a series of tests** you could perform in order to establish the other functional group in **A** and also which of the structures suggested in part **(b)**(ii) is in fact **A**. (6)
(ii) The mass spectrum of **A** includes major peaks at *m/e* values of 15, 45 and 75. Show how this data is consistent with only one of the structures for **A**. (4)

(Total: 25) **Edexcel Specimen Question**

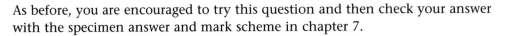

As before, you are encouraged to try this question and then check your answer with the specimen answer and mark scheme in chapter 7.

There are strong connections between many of the key areas or themes of chemistry. These strong connections include links between the particles (atoms, ions, molecules), structure, bonding, properties and uses of materials, those between rates, equilibria and industrial processes and the inter-relationship of the electrochemical (reactivity) series and the periodic table.

These interconnections make chemistry particularly suitable for synoptic assessment which aims to draw together the different areas of a subject.

CHAPTER THREE

ASSESSING SYNOPTIC SKILLS

As part of their assessment of A level Chemistry courses, the awarding bodies are obliged to follow four prescribed assessment objectives.

Assessment Objective 1	**(AO1)**	**Knowledge with Understanding**
Assessment Objective 2	**(AO2)**	**Application of Knowledge and Understanding, Analysis and Evaluation**
Assessment Objective 3	**(AO3)**	**Experiment and Investigation**
Assessment Objective 4	**(AO4)**	**Synthesis of Knowledge, Understanding and Skills**

Assessment Objectives AO1, AO2 and AO3 are the same for Advanced Subsidiary (AS level) and Advanced (A level) GCE courses. Assessment Objective AO4 applies only to the A2 part of the course and is entirely devoted to the assessment of synoptic skills.

There are *small* variations in the percentages of marks which the different chemistry specifications allocate to the first three Assessment Objectives (AO1, AO2 and AO3), but the percentage of marks devoted to Assessment Objective 4 (AO4, effectively synoptic skills) is 20% in all five specifications and this is covered entirely in the A2 course.

The precise way in which synoptic skills are assessed differs from one specification to another. The next part of this chapter therefore contains five tables, one for each of the five specifications – AQA, Edexcel, Edexcel (Nuffield), OCR and OCR (Salters). Each of these tables shows *when* (at what stage of the A2 course), *where* (in which assessment unit or unit test) and *how* (what style of questions, etc.) synoptic skills are assessed. The tables also show the subject content assessed in the various synoptic components, the percentage of the A2 marks, the total A level marks allocated to each assessment unit/unit test and the proportions of each assessment unit/unit test devoted to synoptic skills. Before looking at the table showing the arrangements for the specification which you are studying, remember the following points:

★ synoptic skills account for 20% of the total A level marks in all five specifications and all the synoptic assessment occurs during the A2 course – this means that 40% of the marks in A2 assessment units/unit tests are devoted to synoptic assessment;

★ structured questions involving (i) the planning of experiments, (ii) the use, analysis and interpretation of data and (iii) the prediction of properties, reactions and changes are by far the most common style of synoptic question;

★ synoptic questions will require you to make connections between two or more topic areas and to use your knowledge, understanding and skills in new contexts.

AQA

Assessment Unit	Time	Synoptic question style(s) and relevant subject content	Percentage of A2 marks	Percentage of A level marks
5 Available in January and June	2 hours	Structured questions (120 raw marks), some requiring extended prose All compulsory Content assessed: emphasis on Thermodynamics, Periodicity, Redox Equilibria, Transition Metals, Reactions of Inorganic Compounds in Aqueous Solution.	40 (½ synoptic)	20
6(a) Available in January and June	1 hour	Objective questions (60 raw marks), in the form of multiple choice and multiple completion items All compulsory Content assessed: whole course	20 (all synoptic)	10

Edexcel

Unit Test	Time	Synoptic question style(s) and relevant subject content	Percentage of A2 marks	Percentage of A level marks
5 Available in January and June	1½ hours	<u>Section A</u> Structured questions (50 raw marks) All compulsory <u>Section B</u> Structured questions (25 raw marks) Requiring extended prose Small choice of questions <u>Content assessed</u>: Redox Equilibria, Transition Metal Chemistry, Organic Chemistry III (reaction mechanisms and aromatic compounds), Chemical Kinetics (II), Organic Chemistry (IV) (analysis, synthesis and applications)	20 (½ synoptic) 10 (all synoptic)	10 5
6B Available in January and June	1½ hours	<u>Section A</u> Structured question(s) Requiring extended prose Compulsory Assesses ability to interpret data from laboratory situations <u>Section B</u> Structured questions Requiring extended prose Small choice of questions Assesses ability to apply knowledge and understanding to new contexts Section A & B – 50 raw marks	20 (all synoptic)	10

Edexcel (Nuffield)

Unit Test	Time	Synoptic question style(s) and relevant subject content	Percentage of A2 marks	Percentage of A level marks
6 Available in June	2 hours	Structured questions (60 raw marks) Some requiring extended prose All compulsory Open book test – use of Nuffield A level Chemistry Student's Book (4th Edn) and Nuffield Advanced Science-Book of Data allowed <u>Content assessed</u>: emphasis on The Born–Haber Cycle, Structure & Bonding, Redox Equilibria, Natural & Synthetic Polymers, The Transition Elements, Organic Synthesis, Instrumental Methods	40 (all synoptic)	20

OCR (Salters)

Assessment Unit	Time	Synoptic question style(s) and relevant subject content	Percentage of A2 marks	Percentage of A level marks
2854 Chemistry by Design Available in January and June	2 hours	Structured questions (120 raw marks) Some requiring extended prose All compulsory <u>Content assessed</u>: Aspects of Agriculture, Colour by Design, The Oceans, Medicines by Design	40 (¾ synoptic)	20
2855 Available in January and June	15 to 20 hours work in laboratory	Individual Investigation, Coursework (45 raw marks) Assessment of experimental skills involved in planning experimental work and in the analysis and evaluation of results and data contribute to synoptic assessment	30 (⅓ synoptic)	15

OCR

Assessment Unit	Time	Synoptic question style(s) and relevant subject content	Percentage of A2 marks	Percentage of A level marks
2815 component 01 Available in January and June	1 hour	Structured questions and directed essays (45 raw marks) All compulsory <u>Content assessed</u>: Lattice Enthalpy & Born–Haber Cycles, Periodic Trends of Oxides & Chlorides, Transition Elements & Redox Chemistry	15 (⅔ synoptic)	7.5
2816 component 01 Available in January and June	1¼ hours	Structured questions and directed essays (60 raw marks) All compulsory <u>Content assessed</u>: All the compulsory modules: 2811, 2812, 2813 (component 01), 2814, 2815 (component 01) & 2816 (component 01)	20 (all synoptic)	10
2816 component 02 *or* component 03 Available in January and June	between 5 & 10 hours 1½ hours	Coursework *or* Practical Examination Assessment of experimental and investigative skills (30 raw marks) Synoptic assessments in these components allow students to apply knowledge and understanding in planning experimental work and in the analysis and evaluation of data	20 (½ synoptic)	10

CHAPTER FOUR

DEVELOPING SYNOPTIC SKILLS

Having successfully completed GCSEs, you are, like most A level students, probably adept at learning facts (recall). You will, however, need to *develop* your synoptic skills by making connections between these facts and ideas.

Before looking at ways to develop your synoptic skills, it is well worthwhile listing the key themes which form the framework and structure to Advanced Subsidiary (AS level) and Advanced GCE (A level) Chemistry courses. These key themes are the following.

★ The structure of substances from atomic structure, through electronic structure to molecular, metallic and ionic bonding and crystal structure.

★ The key reactions of inorganic compounds: redox, acid–base, precipitation and complexing.

★ The key reactions of organic compounds: substitution, addition and elimination.

★ The reactivity (electrochemical) series and the periodic table in summarising and rationalising the properties and reactions of all elements.

★ The way in which bonding and structure dictate the properties of a substance and these, in turn, determine its uses.

★ The links between reaction rates and equilibria and the influence of external factors (temperature, concentration, pressure, surface area, catalysts) on both of these and on the choice of conditions for industrial processes.

More than other areas of assessment, synoptic questions will put a heavy emphasis on your understanding and appreciation of key ideas and concepts rather than on your recall of factual information.

If you can acquire a really thorough understanding of the topics in your chemistry course, the burden of assimilating and learning facts will be so much easier and in many situations the facts will simply slot into place. Most of chemistry is an intensely logical study and you should always place a heavy premium on ensuring that you understand key ideas and principles.

In addition to this emphasis on understanding rather than recall, it is important to appreciate the differences between AS and A2 patterns of assessment.

A2 will involve:

- ★ more discussion of key points;
- ★ more explanation and reasoning;
- ★ more predictions of what might happen in novel situations;
- ★ less straightforward description;
- ★ calculations with three or four stages rather than one or two.

Brainstorming and spider diagrams – making connections

Most A level Chemistry courses are taught unit-by-unit, topic-by-topic and some of them are examined in the same way. Although this approach is understandable, it does not allow you to appreciate the connections between different units and topics easily.

Brainstorming, summarised in spider diagrams, will help you to find and appreciate the connections between different areas of your A or AS level courses. Traditionally, 'brainstorming' is the technique of finding solutions to problems or developing new approaches by spontaneous suggestions.

When you brainstorm on a particular topic, such as transition elements or redox, you should produce a list of ideas, concepts and facts associated with the topic under consideration. Later, all this information can be grouped into themes and the various facts and ideas linked by making a 'spider diagram' (figure 4.1).

Using this brainstorming approach, it is very unlikely that the ideas and facts that you have recalled will be confined to a single unit or topic in the AS or A level specification. So, you will be thinking synoptically and 'making connections'. This brainstorming approach can be extended amongst your friends and fellow students by 'playing the game' together or by swapping spider diagrams.

To get the most from brainstorming, you should pick a topic or a theme from within your A2 specification such as Transition Metals (figure 4.1), Periodicity, Equilibria, Reaction Rates (Kinetics), Born–Haber Cycles or Organic Syntheses. In some cases, your specification will identify specific topics for synoptic assessment and it would be wise to include these in your brainstorming sessions. These specific topics for synoptic assessment are listed in the tables in chapter 3.

During brainstorming, it may help to look at the specification for your course and it will certainly help to look at a good generic A level textbook which will make some of the key connections between topics.

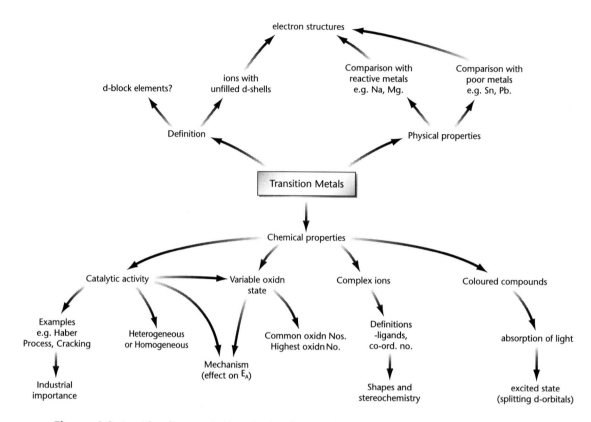

Figure 4.1 A spider diagram linking the key facts and ideas related to transition metals

Two further spider diagrams covering 'Equilibria' and 'Organic Syntheses' are shown in figures 4.2 and 4.3, on pages 23 and 24 respectively.

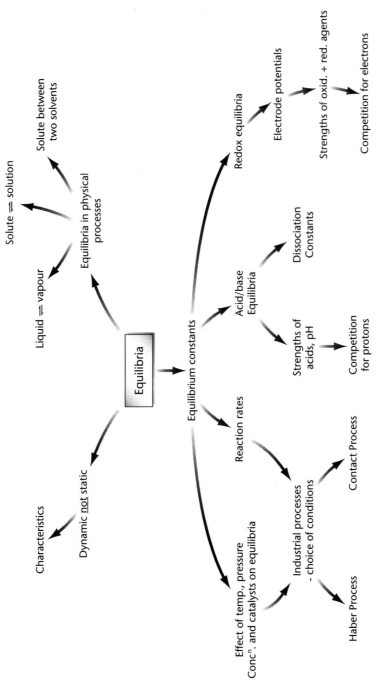

Figure 4.2 A spider diagram related to equilibria

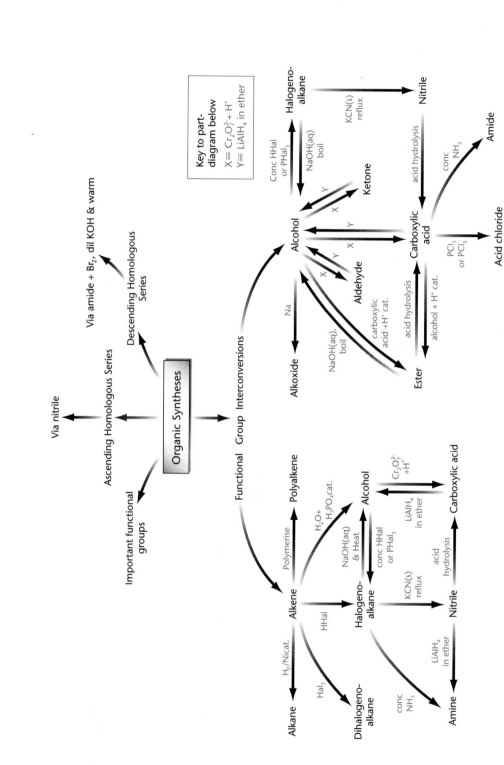

Figure 4.3 A spider diagram related to Organic Syntheses (In this case, it is useful to indicate the chemicals and conditions required for specific interconversions.)

EXAMPLE C

Carry out brainstorms and produce spider diagrams related to

a) Born–Haber cycles

b) Periodicity

There are, of course, no 'right' answers to spider diagrams. Possible answers to each of the two brainstorms in Example C are provided in chapter 7.

As an extension to the spider diagram in figure 4.3 and in order to have a complete summary of functional group interconversions, you should carry out the following exercise in Example D.

EXAMPLE D

Copy out the flow sheet below showing the functional group interconversions for important arenes and write the reagents and conditions alongside the arrow for each conversion.

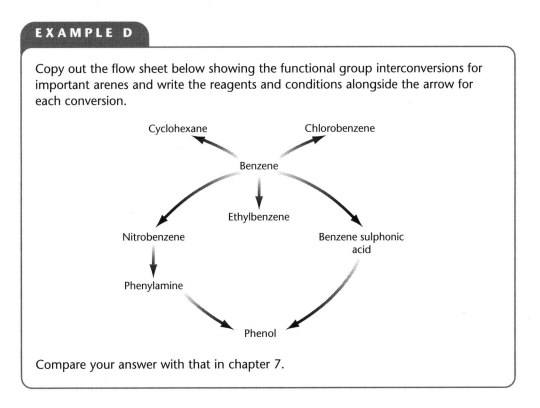

Compare your answer with that in chapter 7.

CHAPTER FIVE

ANALYSING SYNOPTIC SKILLS

In this chapter we shall look at the important skills tested by synoptic questions. These can be divided into four categories:

(i) **Investigative skills** – planning experiments, constructing tables and plotting graphs.

(ii) **Analysing and interpreting data**

(iii) **Predicting** – predicting properties, reactions, changes, etc.

(iv) **Writing directed essays**

(i) Investigative skills

Investigative and experimental work is often synoptic because it requires you to bring together ideas and experiences from various parts of your chemistry studies in order to plan experiments and present your results in tables and graphs.

Here is a very good example of what might be expected of you in planning an experiment. Try answering the questions and then check your answers with the mark scheme in chapter 7.

EXAMPLE E

Plan an experiment to find the percentage by mass of bromine in a bromoalkane. The bromoalkane (b.pt. 91°C) can be hydrolysed completely by refluxing with aqueous sodium hydroxide solution for 45 minutes. The bromide ions produced can be estimated by precipitating them as silver bromide and then weighing the clean, dry precipitate.

a) Write equations for the reactions which occur (i) during refluxing, (ii) during precipitation.   *(2)*

b) How would you carry out the hydrolysis? Describe the apparatus and conditions you would use.   *(4)*

c) Describe how you would obtain a precipitate of silver bromide from the hydrolysed mixture and then determine the mass of clean, dry silver bromide.   *(7)*

d) Show how you would calculate the percentage by mass of bromine in the original bromoalkane.   *(2)*

(Total: 15)

The results of any experimental work should be recorded as the investigation occurs. Your initial results should be presented in a fashion which can be easily understood and then used to draw conclusions.

The usual way of summarising experimental results is to construct a table. The headings for each of the columns in a table should indicate clearly what measurements have been made with the units of these measurements separated by a solidus (forward slash).

Table 5.1 is an example of a results table for an investigation of the rate of the reaction between bromine and methanoic acid.

Time /s	Concentration of bromine $[Br_2(aq)]$/mol dm^{-3}		
	Experiment 1	Experiment 2	Average
0	0.0100	0.0100	0.0100
30	0.0091	0.0089	0.0090
60	0.0081	0.0080	0.0081
90	0.0074	0.0072	0.0073
120	0.0066	0.0065	0.0066
180	0.0057	0.0053	0.0055
240	0.0045	0.0042	0.0044
360	0.0030	0.0026	0.0028
480	0.0021	0.0018	0.0020
600	0.0015	0.0011	0.0013

Table 5.1

It is unlikely that you will be asked to construct a table as part of a written examination, but this may well form part of the synoptic marks for coursework in the OCR and OCR (Salters) specifications.

Most experiments generate quantitative results which can be plotted on a graph to show any trends or patterns most clearly. The results in table 5.1 have been plotted on a graph in figure 5.1.

All graphs should be plotted on graph paper and the mnemonic *ALSUP* will help you to remember the important features which should be included. These features are shown on figure 5.1.

★ **A Axes** should be the correct way round. The horizontal *x* axis is normally used for the independent variable, that is the one which you or the person carrying out the experiment can choose or manipulate (e.g. time in the investigation summarised in table 5.1).

The vertical *y* axis is normally used for the dependent variable, that is the one which is observed or measured (e.g. concentration of bromine in the results in table 5.1).

★ *L* **Labels** should indicate clearly and succinctly the variables being plotted along each axis.

★ *S* **Scales** along the axes should be chosen to show the variation in both vertical (dependent variable) and horizontal (independent variable) changes sufficiently.

★ *U* **Units** for the two variables being plotted should be written close to the labels at the sides of the axes and separated from the labels by a forward slash.

★ *P* **Points** should be plotted accurately and then summarised in a 'line or curve of best fit'.

Points should be shown as crosses (+ or × on the graph paper) and not as dots that would be obscured if the line of best fit passes through the points.

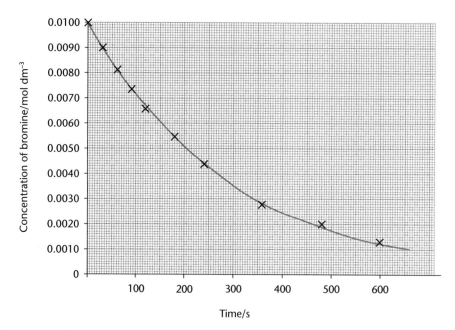

Figure 5.1 The concentration of bromine plotted against time for an investigation of the rate of the reaction between bromine and methanoic acid

Unless you are specifically asked to do so, your line or curve should not be extended or extrapolated beyond the range of your own measurements. Only the measurements or values obtained in your experiment should be represented in your graph. Don't fall into the trap of extrapolating your line or curve to zero.

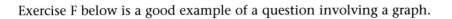

Exercise F below is a good example of a question involving a graph.

EXAMPLE F

The table below shows the concentrations of 2-chloro-2-methylpropane at various times in an experiment to find its order of reaction with sodium hydroxide at 25°C.

Time/min	0	7	15	27	44	60
Conc. of 2-chloro-2-methylpropane/mol dm^{-3}	0.080	0.065	0.054	0.041	0.028	0.020

Show by means of a graph that the reaction is first order with respect to 2-chloro-2-methylpropane.   *(5)*

(ii) *Analysing and interpreting data*

Most synoptic questions will ask you to use, analyse or interpret data in some way or other. You might be asked to describe the results (data) of an experiment, to extract information from a table or a graph, to compare the results of experiments under different conditions, to explain a set of data or to carry out a calculation.

There are different types of questions related to the use of data and these different types can be identified by their opening word (injunction). The main injunctions are 'describe', 'discuss', 'calculate', 'explain' and 'compare', and these are considered separately below.

DESCRIBE This injunction requires you to give a statement or equation summarising the key points, patterns or trends in data you are asked to recall or in data supplied for you as text, in a table or in a graph. Your description should not refer to every measurement or observation, but it should identify any significant points or changes in pattern. In addition, you should point out any anomalous (irregular or unexpected) results that do not fit the general pattern or the usual trends.

Purely 'describe' type questions are, however, rare in synoptic assessment. They will normally be associated with other requirements, such as an explanation or a prediction.

Example G below is the sort of thing you might expect from a 'describe' question.

As before, you should attempt the question and then check your work with the suggested answers in chapter 7.

EXAMPLE G

The table below relates to oxides of Period 3 in the Periodic Table.

Oxide	Na_2O	MgO	Al_2O_3	SiO_2	P_4O_{10}	SO_3
Melting point/°C	1275	2827	2017	1607	580	33
Bonding						
Structure						

a) Copy and complete the table using the following guidelines.

(i) Complete the 'bonding' row using only the words: *ionic* or *covalent*.
(ii) Complete the 'structure' row using only the words: *simple molecular* or *giant*.
(iii) Explain, in terms of forces, the difference between the melting points of MgO and SO_3. *(5)*

b) The oxides Na_2O and SO_3 were each added separately to water. For each oxide, construct a balanced equation for its reaction with water.

(i) SO_3 reaction with water
(ii) Na_2O reaction with water *(2)*

(Total: 7) **OCR Specimen Question**

DISCUSS The injunction 'Discuss' requires a similar approach to that for 'Describe', with a requirement to identify the key points, patterns or opinions on an issue rather than in data.

CALCULATE This injunction requires you to carry out a calculation using data provided in the text of the question, in a table or from a graph. In some cases, you will be expected to look up the data required for your calculation in a specified data book.

In your calculation, you must explain and show your working, giving the answer to the correct number of significant figures, with the appropriate units.

Example H is an example of a 'calculate' question.

EXAMPLE H

The reaction $2SO_2(g) + O_2(g) \rightarrow 2SO_3(g)$ is used industrially for the production of sulphur trioxide in the Contact Process.

Use a data book to calculate an accurate value for the enthalpy change of the reaction.
(The enthalpy change of formation of $SO_3(g)$, $\Delta H_f^\ominus (SO_3(g))$, is $-381\,kJmol^{-1}$.) *(4)*

EXPLAIN The injunction 'Explain' requires you to give reasons for a suggestion or conclusion you may have made or to give reasons for the trends and patterns in data.

Example I is typical of a predominantly 'explain' type question.

EXAMPLE I

A solution of cobalt(II) chloride was reacted with ammonia and ammonium chloride while a current of air was blown through the mixture. A red compound, **X**, was produced which contained a complex ion of cobalt. The compound had the following composition.

	% by mass
Co	23.6
N	27.9
H	6.0
Cl	42.5

a) Use the data to calculate the empirical formula of the compound **X**. (3)

b) In the reaction in which **X** is formed from cobalt(II) chloride, explain the role of:

   (i) the air
   (ii) the ammonium chloride (3)

c) A solution of cobalt(II) chloride reacts with concentrated hydrochloric acid to form a stable complex ion. A solution of calcium chloride does not form a corresponding complex. Use your knowledge of atomic structure to explain this difference, and suggest other differences you would expect between the chemistry of cobalt and calcium. (4)

(Total: 10) *Edexcel (Nuffield) Specimen Question*

COMPARE This requires you to point out similarities *and* differences between two observations or two pieces of data.

A related injunction to 'compare' is 'distinguish between'. In this case, you would be expected to point out the differences but *not* the similarities.

Example J provides assessment using a 'compare' question. Try to answer the question and then compare your answer with the mark scheme in chapter 7.

EXAMPLE J

a) Use a data book to tabulate the electrode systems and their $E^\ominus$ values involving the following elements acting as oxidising agents:
$Br_2(aq)$, $Cl_2(aq)$, $I_2(aq)$ *(1)*

b) When chlorine reacts directly with iron, iron(III) chloride results. Use $E^\ominus$ data to compare this reaction with the reaction in which iron reacts with iodine. *(5)*

(iii) Predicting

Very often the injunctions 'predict' and 'suggest' are used in synoptic questions. The most common situations ask you to predict equations, properties and reactions or suggest changes which will take place. In these questions you are expected to use your knowledge and understanding of chemistry to give an appropriate answer possibly to a novel or unfamiliar situation. In some cases, there may be no right answer and an intelligent chemically-correct answer will earn full marks.

Example K is an excellent example of a 'predict' question.

EXAMPLE K

The structure of the silkworm moth sex attractant, bombykol, is
H_3C—$(CH_2)_2$—CH=CH—CH=CH—$(CH_2)_8$—CH_2OH

a) Predict some of the properties you would expect for bombykol.
You should comment on:

(i) the likely solubility of bombykol in water;
(ii) the number of possible geometric isomers;
(iii) its likely reaction with four reactants of your choice.
Write equations or reaction schemes for the reactions you choose, showing the structures of the organic products clearly. *(8)*

b) When bombykol is treated with ozone in the presence of water, the molecule splits into fragments wherever there is a C=C double bond.

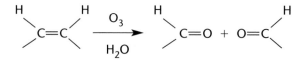

The aldehyde groups are converted to carboxylic acids in oxidizing conditions. Predict the oxidation products formed when bombykol is reacted with ozone in oxidising conditions. *(2)*

(Total: 10) **Edexcel (Nuffield) Specimen Question**

(iv) Writing directed essays

Directed essays are very appropriate for synoptic assessment because they require you to bring together information and ideas from different areas of chemistry. They also require you to organise and present information, ideas and arguments clearly and logically, taking care in your use of grammar, punctuation and spelling.

Synoptic essays test your ability to explain chemical processes and to apply your understanding of chemical principles. In some cases, you will also be expected to show your awareness and appreciation of the applications and implications of chemistry and the chemical industry.

Tips for essay writing.

★ Read the question carefully and answer the question as precisely as possible. Do all that you have been asked to do, but *only* that.

★ Before you start the essay, jot down the key points for your answer.

★ Then, organise your jottings into a logical sequence to form an essay plan.

★ Check that your plan covers all the requirements of the question. For example, if the essay asks for examples from both organic and inorganic chemistry, make sure that these are given equal emphasis.

★ When you have completed the essay, check that you have covered all the requirements of the question by ticking them off on the question paper, and check that you have written legibly in sentences and paragraphs, using good grammar and accurate spelling.

Example L is a good example of a directed essay. Try answering the question and then check your work with the specimen answer and mark scheme in chapter 7. Example L was part of a 2-hour (i.e. 120-minute) specimen paper for which the total possible marks amounted to 120. This works out at a possible acquisition of 1 mark per minute, which implies that you should spend about 15 minutes on the essay.

EXAMPLE L

Write an essay on ammonia.

Your answer should cover the bonding in, and structure of the ammonia molecule, the industrial manufacture of ammonia and at least three of its reactions, including at least one example from both inorganic and organic chemistry. *(15)*

CHAPTER SIX

PRACTISING SYNOPTIC QUESTIONS

This chapter contains five mock examination papers to practise your synoptic skills. The papers are presented and written in the format typical of that used by the awarding bodies.

The different awarding bodies use different names for their examination papers. Edexcel refers to the examination papers as unit tests, whilst AQA and OCR refer to them as assessment units.

For convenience, none of the five papers is more than one hour in length. After completing a paper, preferably under exam conditions, you should check your work against the mark schemes in chapter 7 to see how well you have performed.

Use the mark schemes to correct your work and study the examiner's tips very carefully. Remember 'practice makes perfect'.

Paper 1 (Unit Test 1)

contains 20 objective questions. Although these are used by only AQA, they are an excellent way of revising a wide range of topics from any specification in a relatively short time.

Paper 2 (Unit Test 2)

contains four typical structured questions.

Paper 3 (Unit Test 3)

contains a further three structured questions similar in general style to those in paper 2.

Paper 4 (Unit Test 4)

contains three structured questions which reflect laboratory situations plus the interpretation and analysis of results.

Paper 5 (Unit Test 5)

contains four directed essays.

SURNAME		INITIALS
SIGNATURE		

CENTRE NO.

CANDIDATE NO.

AB1 AWARDING BODY 1
Advanced GCE

SCIENCE: CHEMISTRY
Paper 1 Synoptic Assessment

In addition to this paper you will require:

- a Periodic Table (see page 56)
- a blue or black pen
- you may use a calculator.

Time allowed: 30 minutes

Instructions

Use blue or black ink.

Fill in the boxes at the top of this page.

Answer **all** 20 questions.

For each question there are four responses – A, B, C and D. When you have selected the response which you think is the best answer to a question, write the appropriate letter at the right hand side of the question number in the Answer Grid.

If you wish to change your answer to a question, cross out your initial response neatly and write your preferred answer alongside.

Do all rough work on additional paper. Cross through any work you do not want marked.

Information

Each correct answer will score one mark. No deductions will be made for wrong answers.

This paper carries 10 per cent of the total marks for Advanced Level.

The following data may be required:

Gas constant $R = 8.31 \, J \, mol^{-1} \, K^{-1}$

Advice

Do not spend too long on any question. If you have time at the end, go back and answer any question you missed out.

ANSWER GRID	
Question	**Answer**
1	
2	
3	
4	
5	
6	
7	
8	
9	
10	
11	
12	
13	
14	
15	
16	
17	
18	
19	
20	

Multiple Choice Questions

Each of questions 1 to 10 is followed by four responses, A, B, C and D.

For each question select the best response and mark its letter on the answer sheet.

1. Which *one* of the following electronic structures is that of an element with a maximum oxidation number of +3?

   A $1s^2 2s^2 2p^3$

   B $1s^2 2s^2 2p^6 3s^2 3p^1$

   C $1s^2 2s^2 2p^6 3s^2 3p^3$

   D $1s^2 2s^2 2p^6 3s^2 3p^6 3d^3 4s^2$

2. Which *one* of the following has an overall dipole moment of zero?

   A CH_2Cl_2 B NCl_3 C H_3O^+ D NH_4^+

3. Hydrogen peroxide is oxidised by acidified potassium manganate(VII) according to the following half-equations.

$$H_2O_2 \rightarrow O_2 + 2H^+ + 2e^-$$
$$MnO_4^- + 8H^+ + 5e^- \rightarrow Mn^{2+} + 4H_2O$$

   What volume of 0.02 mol dm^{-3} KMnO$_4$ is required to oxidise 0.002 mol of H_2O_2?

   A $10\ cm^3$ B $20\ cm^3$ C $40\ cm^3$ D $50\ cm^3$

4. Which *one* of the following reactions involves a decrease in entropy?

   A $CuCO_3(s) \rightarrow CuO(s) + CO_2(g)$

   B $Zn(s) + H_2SO_4(aq) \rightarrow ZnSO_4(aq) + H_2(g)$

   C $[Cu(NH_3)_4]^{2+}(aq) + edta^{4-}(aq) \rightarrow [Cu(edta)]^{2-}(aq) + 4NH_3(aq)$

   D $N_2(g) + 3H_2(g) \rightarrow 2NH_3(g)$

5. Which *one* of the following is a true statement?

   A Across period 3 from Na to Cl, the atomic radius decreases.

   B Across the d-block elements from Sc to Zn, the atomic radius increases.

   C Down group 2 from Mg to Ba, the $E^\ominus$ values for $M^{2+} \rightarrow M$ become more positive.

   D Down group 7 from Cl to I, the ionisation energy increases.

6. Increasing the temperature of the reaction

$$X(g) + Y(g) \rightarrow Z(g) \qquad \Delta H = -x \text{ kJ}$$

  A decreases the enthalpy change of the reaction.

  B increases the rate constant of the reaction.

  C reduces the activation energy.

  D increases the equilibrium constant.

7. Which *one* of the following shows two molecules with the same molecular shapes relative to their constituent atoms?

  A BF_3 and NH_3

  B $BeCl_2$ and HCN

  C H_2O and HCN

  D SO_2 and CO_2

8. Which *one* of the following compounds has a stereoisomer?

  A $Co(NH_3)_5Cl$

  B $CHCl{=}CCl_2$

  C $CH_3CHClCH_2CH_3$

  D $CH_3CH_2CONH_2$

9. Which *one* of the following reactions does not produce ethanoic acid?

  A warming CH_3CHO with dilute acidified potassium dichromate(VI)

  B warming CH_3COOCH_3 with a little conc. sulphuric acid

  C warming CH_3COONa with conc. sulphuric acid

  D warming CH_3COCH_3 with dilute acidified potassium manganate(VII)

10. Which *one* of the following is a redox reaction?

  A $CH_4 + Cl_2 \rightarrow CH_3Cl + HCl$

  B $BaCl_2 + Na_2CrO_4 \rightarrow BaCrO_4 + 2NaCl$

  C $CH_3COOH + NaOH \rightarrow CH_3COONa + H_2O$

  D $Al(OH)_3 + NaOH \rightarrow NaAl(OH)_4$

Multiple Completion Questions

In each of questions 11 to 20, one or more of the responses is/are correct. Decide which of the responses to the question is/are correct and mark **A**, **B**, **C** or **D** on the answer sheet as follows.

A if (i), (ii) and (iii) only are correct.

B if (i) and (iii) only are correct.

C if (ii) and (iv) only are correct.

D if (iv) alone is correct.

Summarised directions for recording responses to multiple completion questions			
A	**B**	**C**	**D**
(i), (ii) and (iii) only	(i) and (iii) only	(ii) and (iv) only	(iv) alone

11. Which of the following is/are oxidised by warming with a dilute solution of acidified potassium manganate(VII)?

    (i) Fe^{3+} **(ii)** Cu^{2+}

    (iii) CH_3COCH_3 **(iv)** CH_3CH_2CHO

12. Which of the following is/are hydrogen bonded in the liquid state?

    (i) CH_3CHO **(ii)** CH_3CH_2OH

    (iii) CH_2F_2 **(iv)** CH_3NH_2

13. In the reaction:

$$2H_2O \rightarrow OH^- + H_3O^+$$

water is acting

    (i) as an acid   **(ii)** as a base

    (iii) as a base and an acid   **(iv)** as neither a base nor an acid

14. Which of the following react(s) with water to give an acidic solution?

    (i) CH_3Cl **(ii)** CH_3COCl

    (iii) PH_3 **(iv)** PCl_3

15. In which of the following molecules do all the atoms lie in one plane?

    (i) BF_3 **(ii)** $H—C{\equiv}C—H$

    (iii) ⬡–Cl   **(iv)** PH_3

16. Which of the following has/have the same number of electrons as the chloride ion, Cl^-?

   (i) Ca^{2+}

   (ii) PH_3

   (iii) S^{2-}

   (iv) Na^+

17. Which of the following properties of the halogens increase(s) with increasing relative atomic mass?

   (i) electropositive character

   (ii) strength as oxidising agent

   (iii) absorption of visible light

   (iv) first ionisation energy

18. The following data refers to the reaction

$$2A(g) + B(g) \rightarrow A_2B(g)$$

Expt.	Concentration of A/mol dm^{-3}	Concentration of B/mol dm^{-3}	Initial reaction rate /mol dm^{-3} s^{-1}
1	0.1	0.4	0.1
2	0.2	0.4	0.4
3	0.4	0.4	1.6
4	0.4	0.2	1.6
5	0.4	0.1	1.6

   Which of the following conclusions can be drawn from the data.

   (i) The reaction is second order with respect to A.

   (ii) The reaction is first order with respect to B.

   (iii) The reaction cannot occur in a single step.

   (iv) The first step in the reaction is $A(g) + B(g) \rightarrow AB(g)$

19. Ethanol, CH_3CH_2OH, and phenol, ⬡-OH, react in a similar way with

   (i) bromine water (Br_2(aq))

   (ii) sodium metal

   (iii) neutral iron(III) chloride solution

   (iv) ethanoyl chloride (CH_3COCl)

20. Ethanol and ethanoic acid produce at least one product in common when they react separately with

   (i) conc. sulphuric acid

   (ii) potassium manganate(VII)

   (iii) sodium hydroxide

   (iv) sodium

Centre No.						Surname		Initial(s)
Candidate No.						Signature		

AWARDING BODY 2 GCE

Science: Chemistry

Paper 2 (Unit Test 2)

Advanced

Time: 1 hour

Materials required for examination

Graph paper
Answer booklet
Periodic Table (see page 56)

Instructions to Candidates

In the boxes above, write your centre number, candidate number, your surname, initials and signature.

Answer ALL the questions.

Show all the stages in any calculations and state the units. Calculators may be used.

Final answers to calculations should be given to an appropriate number of significant figures.

Include diagrams in your answers where these are helpful.

Information for Candidates

The marks for the various parts of questions are shown in square brackets: e.g. [2].

This paper has four questions. There are no blank pages.

The total mark for this paper is 50.

This test forms part of your synoptic assessment. All the questions in this test are designed to give you the opportunity to 'make connections between different areas of chemistry and to use skills and ideas developed throughout the course in new contexts'. You should include in your answers relevant information from the whole of your course, where appropriate.

Advice to Candidates

You are reminded of the need to organise and present information, ideas, descriptions and arguments clearly and logically, taking account of your use of grammar, punctuation and spelling.

1. The pH of blood is maintained in healthy individuals by various buffering systems. One of the most important systems contains carbon dioxide, CO_2, and hydrogencarbonate ions, HCO_3^-, linked by the reaction:

   $$CO_2(aq) + H_2O(l) \rightleftharpoons H^+(aq) + HCO_3^-(aq) \quad \textbf{Equation 1}$$

   K_a for this reaction has the value of 4.5×10^{-7} mol dm^{-3} at the temperature in the body.

   (a) (i) Give the mathematical definition of pH. [1]

   (ii) Write the expression for K_a for the equilibrium shown in **Equation 1** in terms of the concentrations of $CO_2(aq)$, $H^+(aq)$ and $HCO_3^-(aq)$. [2]

   (b) In the blood of a healthy person:

   $$[HCO_3^-(aq)] = 2.5 \times 10^{-2} \text{ mol dm}^{-3}$$

   $$[CO_2(aq)] = 1.25 \times 10^{-3} \text{ mol dm}^{-3}$$

   (i) Use this data, your answer to **(a) (ii)** and the value of K_a to calculate the concentration of H^+ ions in the blood of a healthy person. [2]

   (ii) Calculate the pH of the blood of a healthy person. [1]

   (c) Use **Equation 1** and the information in **(b)** to explain how blood acts as a buffer solution.

   $$CO_2(aq) + H_2O(l) \rightleftharpoons H^+(aq) + HCO_3^-(aq) \quad \textbf{Equation 1}$$

   [5]

   Note: In this question 1 mark is available for the quality of written communication.

   (d) The equilibrium reaction in **Equation 1** plays a dual role in controlling the acidity of the oceans and controlling the concentration of carbon dioxide in the atmosphere.

   Write two further equations, with state symbols, to show how dissolved carbon dioxide in the oceans can be converted to calcium carbonate in sea shells. [2]

   (OCR(Salters) – Specimen question) **Total [13]**

2. Hydrogen reacts with iodine at 450°C to give hydrogen iodide. The results from several experiments designed to find the rate equation for the reaction are given below.

Initial $[I_2]$ /mol dm^{-3}	Inital $[H_2]$ /mol dm^{-3}	Relative initial rate
0.001	0.001	1
0.003	0.001	3
0.001	0.004	4

(a) (i) Find the order of reaction with respect to each of the reactants.

[3]

(ii) Write the rate equation for the reaction. [1]

(iii) Explain why rate equations cannot be written from the stoichiometric (chemical) equation for the reaction, but must be found experimentally. [2]

(iv) What do you understand by the **rate-determining step** in a chemical reaction? [1]

(v) Illustrate your answer to (iv) by writing out any reaction mechanism of your choice. [3]

(b) Reaction rates generally increase with an increase in temperature for two reasons.

(i) State what these reasons are. [2]

(ii) Copy and then sketch on the axes a graph of the distribution of molecular energies at a given temperature T_1 and at a higher temperature T_2, and hence explain the increase in the rate.

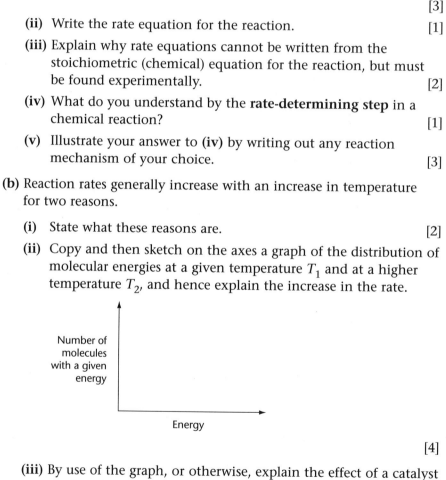

[4]

(iii) By use of the graph, or otherwise, explain the effect of a catalyst on the rate of a chemical reaction. [2]

(Edexcel – Specimen question) **Total [18]**

3. Sodium bromide is formed from its elements at 298 K according to the equation

$$Na(s) \ + \ \tfrac{1}{2}Br_2(l) \ \rightarrow \ NaBr(s)$$

The lattice dissociation enthalpy of solid sodium bromide refers to the enthalpy change for the process

$$NaBr(s) \ \rightarrow \ Na^+(g) \ + \ Br^-(g)$$

The electron addition enthalpy refers to the process

$$Br(g) \ + \ e^- \ \rightarrow \ Br^-(g)$$

Use this information and the data in the table below to answer the questions which follow.

Standard enthalpies		$\Delta H^\ominus$/kJ mol^{-1}
$\Delta H_f^\ominus$	formation of NaBr(s)	-361
$\Delta H_{ea}^\ominus$	electron addition to Br(g)	-325
$\Delta H_{sub}^\ominus$	sublimation of Na(s)	$+107$
$\Delta H_{diss}^\ominus$	bond dissociation of Br$_2$(g)	$+194$
$\Delta H_i^\ominus$	first ionisation of Na(g)	$+498$
$\Delta H_L^\ominus$	lattice dissociation of NaBr(s)	$+753$

(a) Construct a Born–Haber cycle for sodium bromide. Label the steps in the cycle with symbols like those used above rather than numerical values. [6]

(b) Use the data above and the Born–Haber cycle in part (a) to calculate the enthalpy of vaporisation, $\Delta H_{vap}^\ominus$, of liquid bromine. [3]

(AQA – Specimen question) Total [9]

4.   The structures of the two isomers, **X** and **Y**, are shown below:

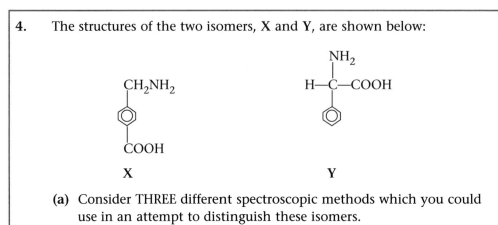

          X   **Y**

(a) Consider THREE different spectroscopic methods which you could use in an attempt to distinguish these isomers.

Indicate the results you would expect, supporting your answer with data where appropriate.   [6]

(b) Only one of the isomers is chiral. Identify which isomer is chiral. State the physical property caused by the chirality which would allow you to distinguish the isomers.   [2]

(c) Isomer **X** can be converted to the following compound, **Z**.

Z

Molecules of compound **Z** can react to form a polymer. Draw a section of the polymer, showing at least two monomer units, and indicate any other products which form.   [2]

(Edexcel (Nuffield) – Specimen question)   Total [10]

Candidate Name	Centre Number	Candidate Number

AB3

AWARDING BODY 3
Advanced GCE

SCIENCE: CHEMISTRY   PAPER 3 (UNIT TEST 3)

Additional materials required:
   Answer paper,
   Periodic Table (see page 56).

TIME   1 hour

INSTRUCTIONS TO CANDIDATES

Write your name, centre number and candidate number in the spaces at the top of this page.

Answer **all** questions.

INFORMATION FOR CANDIDATES

The number of marks is given in brackets [ ] at the end of each question or part question.

You will be awarded marks for the quality of written communication where an answer requires a piece of extended writing.

In this paper you are expected to show your knowledge and understanding of different aspects of Chemistry and the connections between them.

The total mark for this paper is 50.

➤

1. Ethanol, C_2H_5OH, is an important industrial chemical with about 200,000 tonnes manufactured in the UK each year. The usual method of manufacture is by the hydration of ethene with steam in the presence of a phosphoric acid catalyst at 550 K and a pressure of about 7000 kPa.

$$C_2H_4(g) + H_2O(g) \rightleftharpoons C_2H_5OH(g) \quad \Delta H = -46 \text{ kJ mol}^{-1}$$

(a) (i) Predict, with justification, the optimum conditions for this reaction.

   (ii) Explain why the actual conditions used may be different from the optimum conditions.

   (iii) The boiling points of the three chemicals involved in this equilibrium are shown in the table below.

Compound	C_2H_4	H_2O	CH_3CH_2OH
Boiling point/°C	−104	100	78

     Suggest how the ethanol could be separated from the equilibrium mixture. [8]

(b) (i) Write an expression for K_p of this reaction.

   (ii) Explain, with a reason in each case, whether you would expect the value of K_p to alter if any of the external variables below were changed as indicated.

- *increase in temperature*
- *increase in pressure*
- *presence of catalyst* [5]

(c) Alcohols such as ethanol can be used as alternative fuels to petrol. The combustion of ethanol tends to be more complete than the combustion of the alkanes present in petrol, partly because less oxygen is required for combustion.

   (i) Use equations to compare the amount of oxygen required per gram of fuel combusted.

   (ii) Suggest why there is this difference between the amount of oxygen required per gram for these two fuels. [5]

(OCR – Specimen question) **Total [18]**

2. **(a) (i)** Write the electronic configurations for Cu and for Cu^+.

    (ii) In what way is the electronic structure of copper unusual in terms of the general trend across the first transition series?

    (iii) Explain why the copper(I) ion is not coloured. [4]

(b) Use the electrode potential data concerning copper to answer the questions that follow.

$$Cu^{2+}(aq) + e^- \rightarrow Cu^+(aq) \qquad E = +0.15 \text{ V}$$

$$Cu^+(aq) + e^- \rightarrow Cu(s) \qquad E = +0.52 \text{ V}$$

Suggest what would happen if a sample of copper(I) sulphate, Cu_2SO_4, was added to water. State in general terms the nature of the process which is occurring and state what you would see if copper(I) sulphate was added to water. [6]

(c) Copper(II) sulphate solution contains the complex ion $[Cu(H_2O)_6]^{2+}$.

    (i) Draw the 'electrons in boxes' diagram for this complex ion, distinguishing clearly the copper electrons from the ligand electrons.

    (ii) State what you would see if a solution of aqueous ammonia was added dropwise to a solution of this ion. Show by means of equations how this reaction proceeds and state the type of reaction occurring at each stage. [7]

(Edexcel – Specimen question) **Total [17]**

Synoptic Skills in Advanced Chemistry

3. The compound *Mecoprop* is used as a herbicide. It is structurally similar
    to a number of other compounds which are also used as herbicides. One
    structural feature which these molecules share with *Mecoprop* is the
    presence of a chlorinated benzene ring.

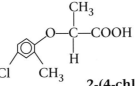

**2-(4-chloro-2-methylphenoxy)propanoic acid
or *Mecoprop***

(a) (i)   State the reagent(s) and conditions which can be used to
           substitute a chlorine atom into a benzene ring.   [3]

    (ii) It is usually necessary to use a catalyst in order to substitute a
          chlorine atom into a benzene ring. Explain the function of the
          catalyst in this process.   [3]

Mecoprop and related herbicides are solids with very low solubilities in
water. They are considerably more soluble in organic solvents such as
ethyl ethanoate. However, drinking water in certain areas of the UK has
been found to be contaminated with traces of *Mecoprop* and other
herbicides. The identities of the herbicides in a sample of water can be
found by first extracting them into a solvent such as ethyl ethanoate,
concentrating the solution formed and then analysing this solution.

(b) Describe, in outline, how you could take a sample of water
    containing dissolved herbicides and extract the herbicides into a
    solvent such as ethyl ethanoate (which is immiscible with, and less
    dense than water), and then concentrate the resulting solution.   [5]

(c) One way to analyse the resulting solution is by gas–liquid
    chromatography (g.l.c.). Before a sample of this solution is injected
    into the column, it is treated with diazomethane to convert any
    carboxylic acid groups present into methyl esters.

    (i)   Suggest why the sample is treated in this way.   [2]

    (ii) Draw out the full structural formula of the methyl ester
          produced from *Mecoprop*.   [2]

(OCR (Salters) – Specimen question)   *Total [15]*

**General Certificate of Education
Advanced Level Examination**

Science: Chemistry

AB4

Paper 4 (Unit Test 4)

Awarding Body 4

Centre number					Candidate number				
Surname									
Other names									
Candidate signature									

In addition to this paper you will need
- a Periodic Table (see page 56)
- access to a book of data

Time
- 1 hour

Instructions to candidates

- Write your name and other details in the spaces provided above.

- Answer **all** questions.

- Write your answers in blue or black ink or ball-point pen.

- All working must be shown. Final answers to calculations should show the units and be given to an appropriate number of significant figures.

- Cross through any work you do not want marked.

For examiner's use	
1	
2	
3	
Total	

Information for candidates

- The number of marks is given in brackets at the end of each question or part-question.

- The maximum mark for this paper is 50.

- You are expected to use a calculator where appropriate.

- You will be assessed on your ability to use an appropriate form and style of writing, to organise relevant information clearly and coherently, and to use specialist vocabulary, where appropriate.

➤

1.   Certain metal oxides will catalyse the decomposition of hydrogen peroxide forming water and oxygen. An equation for the reaction is:

$$2H_2O_2(aq) \rightarrow 2H_2O(l) + O_2(g)$$

This question involves planning an experiment to compare the decomposition of hydrogen peroxide with two metal oxides by measuring the volume of oxygen produced in 2 minutes immediately after hydrogen peroxide solution is added to a catalyst.

(a) Calculate the volume of oxygen produced at 20°C and 100 kPa when 100 cm^3 of 1.00 mol dm^{-3} hydrogen peroxide decomposes. (The gas constant, $R = 8.31$ J mol^{-1} K^{-1}.)   [5]

(b) Use your answer in part (a) to help you choose a sensible concentration and volume of hydrogen peroxide solution, suitable apparatus and an appropriate procedure for the experiment.   [10]

Total [15]

2. Hard water contains dissolved calcium or magnesium ions, which come from the rocks over which the water has flowed. The concentration of these ions can be found in several ways.

(a) One method is to titrate a known volume of the water with a standard solution of the compound edta.

edta complexes with Ca^{2+} ions in a 1:1 ratio. To ensure that the edta complexes satisfactorily with the Ca^{2+} ions, the solution must be buffered at about pH 10. At this pH the indicator used will change colour when all the Ca^{2+} ions have been complexed.

(i) The buffer which is usually used in this titration is a mixture of aqueous ammonia and ammonium chloride solution. Explain how this mixture behaves as a buffer solution. [6]

(ii) A 50.0 cm^3 sample of tap water was titrated with edta solution of concentration 0.0100 mol dm^{-3}. In the titration 31.2 cm^3 of the edta solution was needed before the indicator changed colour. What is the calcium ion concentration in the water in mol dm^{-3}? [3]

(b) A second method for the determination of Ca^{2+} uses the precipitation of the salt calcium ethanedioate, CaC_2O_4. The precipitate is filtered off, dissolved in warm dilute nitric acid, and the ethanedioate, now in solution as ethanedioic acid, is determined by titration with standard potassium manganate(VII) solution.

(i) 25.0 cm^3 of a solution of Ca^{2+} ions containing 0.0500 mol dm^{-3} of Ca^{2+} ions was treated with excess ammonium ethanedioate solution, and the precipitate of calcium ethanedioate was filtered off. Find the mass of the salt which would be precipitated. [3]

(ii) The precipitate was washed with warm dilute nitric acid until completely dissolved, and the washings made up to 250 cm^3 with pure water. 25.0 cm^3 portions of this solution were added to about 25 cm^3 of dilute sulphuric acid, and the mixture was titrated at 60°C with 0.00200 mol dm^{-3} potassium manganate(VII) solution. Write the equation for the reactions occurring during the titration and calculate the volume of potassium manganate(VII) solution which would be required. [5]

(Edexcel – Specimen question)   Total [17]

3.   A student prepared benzoic acid, C_6H_5COOH, by hydrolysing methyl benzoate, $C_6H_5COOCH_3$, using the following method.

- *Dissolve 4.0 g of sodium hydroxide in water to make 50 cm³ of an alkaline solution.*
- *Add the aqueous sodium hydroxide to 2.70 g of methyl benzoate in a 100 cm³ flask and set up the apparatus for reflux.*
- *Reflux this mixture for 30 minutes.*
- *Distil the mixture and collect the first 2 cm³ of distillate.*
- *Pour the residue from the flask into a beaker and add dilute sulphuric acid until the solution is acidic.*
- *Filter the crystals obtained and re-crystallise from hot water to obtain the benzoic acid.*

The overall equation for this hydrolysis is:

$$C_6H_5COOCH_3 + H_2O \rightarrow C_6H_5COOH + CH_3OH$$

The student obtained 1.50 g of benzoic acid, C_6H_5COOH.

(a) Name the functional group that reacts during this hydrolysis.   [1]

(b) (i) Calculate how many moles of methyl benzoate were used.

    (ii) What was the concentration, in mol dm^{-3}, of the aqueous sodium hydroxide used?

    (iii) Calculate the percentage yield of the C_6H_5COOH obtained by the student.

    (iv) Suggest why the percentage yield was substantially below 100%.   [9]

(c) (i) Why was the residue from the flask acidified before recrystallising?

    (ii) Why were the crystals recrystallised?   [2]

(d) Infra-red spectroscopy can be used to monitor the progress of a chemical reaction.

    (i) Predict the key identifying features of the infra-red spectra of methyl benzoate and its hydrolysis products, benzoic acid and methanol.

    (ii) How could you use infra-red spectroscopy to show that the methanol did **not** contain any benzoic acid.   [6]

(OCR – Specimen question)   Total [18]

| Centre No. | | | | | | Surname | | Initial(s) |
| Candidate No. | | | | | | Signature | | |

AWARDING BODY 5
GCE

Science: Chemistry

Paper 5

Unit Test 5 (Synoptic)

Time: 1 hour

Materials required for examination

Periodic Table (see page 56)
Book of data

Instructions to Candidates

In the boxes above, write your centre number, candidate number, your surname, initials and signature.

Answer ALL questions.

Show all the stages in any calculations and state the units. Calculators may be used.

Final answers to calculations should be given to an appropriate number of significant figures.

Information for Candidates

The marks for the various parts of questions are shown in square brackets: e.g. [2].

This paper has four questions. The total marks for this paper are 52.

Advice to Candidates

This question paper is designed to give you the opportunity to make connections between different areas of Chemistry and to use skills and ideas developed throughout the course in new contexts. You should include in your answers relevant information from the whole of your course, where appropriate.

You will be assessed on your ability to organise and present information, ideas, descriptions and arguments clearly and logically, taking account of your use of grammar, punctuation and spelling.

Question Number	Leave Blank
1	
2	
3	
4	
Total	

1. Using knowledge, principles and concepts from different areas of chemistry, explain and interpret, as fully as you can, the data given in the table below. In order to gain full credit, you will need to consider each type of information separately and also to link this information together.

    (*In this question, 1 mark is available for the quality of written communication.*)

Compound	Boiling point/K	Properties of a 0.1 mol dm^{-3} solution	
		Electrical conductivity	[H$^+$]/mol dm^{-3}
NaCl	1686	good	1.0×10^{-7}
CH$_3$COOH	391	slight	1.3×10^{-3}
CH$_3$CH$_2$OH	352	poor	1.0×10^{-7}
AlCl$_3$	451	good	3.0×10^{-1}

(OCR – Specimen question)   **Total [14]**

2. Ethane can be cracked at high temperatures to yield ethene and hydrogen, according to the equation:

    $$C_2H_6(g) \rightleftharpoons C_2H_4(g) + H_2(g)$$

    The standard enthalpy of formation of ethene is positive whereas that of ethane is negative.

    (a) Discuss the effect on the equilibrium constant, K_p, of changes to

    (i) the temperature

    (ii) the pressure. [3]

    (b) Calculate the value of the equilibrium constant, K_p, for this cracking reaction, given that 1.00 mol of ethane under equilibrium pressure of 180 kPa at 1000 K can be cracked to produce an equilibrium yield of 0.36 mol of ethene. [7]

    (AQA – Specimen question)   **Total [10]**

3. Redox reactions are an important type of reaction in chemistry.

   Explain what is meant by a redox reaction. Illustrate your answer with **two** examples drawn from inorganic chemistry (one of which should involve a transition element) and **two** examples from organic chemistry. (*In this question, 1 mark is available for the quality of written communication.*)

   (OCR – Specimen question)　*Total [14]*

4. **(a)** When a mixture of chlorine with an excess of methane is irradiated with ultraviolet light, a reaction occurs with chloromethane as the main organic product.

   Write an equation and a mechanism for the formation of chloromethane. [5]

   (b) Ethanol can be produced industrially either by the direct hydration of ethene or by fermentation. State the conditions for each method and compare the two processes by giving two advantages and one disadvantage of direct hydration. [9]

   (AQA – Specimen question)　*Total [14]*

DATA SHEET
The Periodic Table of the Elements

Group																	
I	**II**											**III**	**IV**	**V**	**VI**	**VII**	**0**
				1 **H** Hydrogen 1													4 **He** Helium 2
7 **Li** Lithium 3	9 **Be** Beryllium 4											11 **B** Boron 5	12 **C** Carbon 6	14 **N** Nitrogen 7	16 **O** Oxygen 8	19 **F** Fluorine 9	20 **Ne** Neon 10
23 **Na** Sodium 11	24 **Mg** Magnesium 12											27 **Al** Aluminium 13	28 **Si** Silicon 14	31 **P** Phosphorus 15	32 **S** Sulphur 16	35.5 **Cl** Chlorine 17	40 **Ar** Argon 18
39 **K** Potassium 19	40 **Ca** Calcium 20	45 **Sc** Scandium 21	48 **Ti** Titanium 22	51 **V** Vanadium 23	52 **Cr** Chromium 24	55 **Mn** Manganese 25	56 **Fe** Iron 26	59 **Co** Cobalt 27	59 **Ni** Nickel 28	63.4 **Cu** Copper 29	65.4 **Zn** Zinc 30	70 **Ga** Gallium 31	73 **Ge** Germanium 32	75 **As** Arsenic 33	79 **Se** Selenium 34	80 **Br** Bromine 35	84 **Kr** Krypton 36
85 **Rb** Rubidium 37	88 **Sr** Strontium 38	89 **Y** Yttrium 39	91 **Zr** Zirconium 40	93 **Nb** Niobium 41	96 **Mo** Molybdenum 42	**Tc** Technetium 43	101 **Ru** Ruthenium 44	103 **Rh** Rhodium 45	106 **Pd** Palladium 46	108 **Ag** Silver 47	112 **Cd** Cadmium 48	115 **In** Indium 49	119 **Sn** Tin 50	122 **Sb** Antimony 51	128 **Te** Tellurium 52	127 **I** Iodine 53	131 **Xe** Xenon 54
133 **Cs** Caesium 55	137 **Ba** Barium 56	139 **La** Lanthanum 57 *	178 **Hf** Hafnium 72	181 **Ta** Tantalum 73	184 **W** Tungsten 74	186 **Re** Rhenium 75	190 **Os** Osmium 76	192 **Ir** Iridium 77	195 **Pt** Platinum 78	197 **Au** Gold 79	201 **Hg** Mercury 80	204 **Tl** Thallium 81	207 **Pb** Lead 82	209 **Bi** Bismuth 83	**Po** Polonium 84	**At** Astatine 85	**Rn** Radon 86
Fr Francium 87	226 **Ra** Radium 88	227 **Ac** Actinium 89 †															

*58–71 Lanthanoid series
†90–103 Actinoid series

140 **Ce** cerium 58	141 **Pr** Praseodymium 59	144 **Nd** Neodymium 60	**Pm** Promethium 61	150 **Sm** Samarium 62	152 **Eu** Europium 63	157 **Gd** Gadolinium 64	159 **Tb** Terbium 65	162 **Dy** Dysprosium 66	165 **Ho** Holmium 67	167 **Er** Erbium 68	169 **Tm** Thulium 69	173 **Yb** Ytterbium 70	175 **Lu** Lutetium 71
232 **Th** Thorium 90	**Pa** Protactinium 91	238 **U** Uranium 92	**Np** Neptunium 93	**Pu** Plutonium 94	**Am** Americium 95	**Cm** Curium 96	**Bk** Berkelium 97	**Cf** Californium 98	**Es** Einsteinium 99	**Fm** Fermium 100	**Md** Mendelevium 101	**No** Nobelium 102	**Lr** Lawrencium 103

Key

a **X** b	a = relative atomic mass **X** = atomic symbol b = atomic (proton) number

The volume of one mole of any gas is 24 dm³ at room temperature and pressure (r.t.p.).

CHAPTER SEVEN

MARKING SYNOPTIC QUESTIONS

This chapter provides suggested answers to the examples labelled A to L in the first five chapters, followed by mark schemes and examiner's tips for the five mock examination papers in chapter 6.

In the mark schemes, marks for specific points are shown in round brackets (e.g. (1)).

Sometimes, a section of the text in a mark scheme is underlined to indicate the key point. A second acceptable answer is given in some cases, separated from the first, and perhaps more obvious answer, by a forward slash.

Suggested answers to Examples A to L

Example A

(a) (i) Ratio of masses $Fe : Cl = 44.0 : 56.0$

Ratio of moles $Fe : Cl = \dfrac{44.0 : 56.0}{55.8 \quad 35.5} = 0.79 : 1.58$ (1)

$= 1 : 2$

∴ Compound A is $FeCl_2$/iron(II) chloride   (1)

Examiner's tip
Having calculated the ratio of moles to the appropriate number of significant figures, it is helpful to divide by the smallest in order to obtain a ratio of moles in integral values.

(ii) $Fe + 2HCl \rightarrow FeCl_2 + H_2$ (1)

(b) (i) Ratio of masses $Fe : Cl = 2.79 : 5.33$ (1)

Ratio of moles $Fe : Cl = \dfrac{2.79 : 5.33}{55.8 \quad 35.5} = 0.050 : 0.150$ (1)

$= 1 : 3$

∴ Compound B is $FeCl_3$/iron(III) chloride   (1)

(ii) $2Fe + 3Cl_2 \rightarrow 2FeCl_3$ (1)

➤

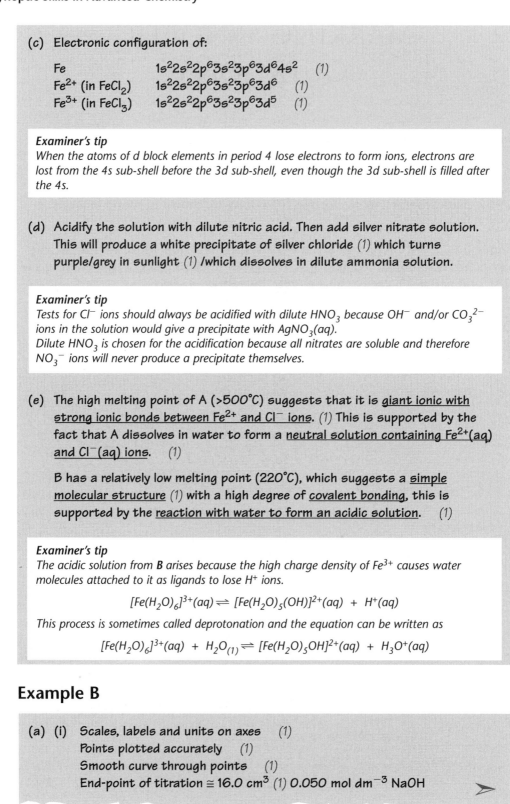

(c) Electronic configuration of:

Fe $1s^2 2s^2 2p^6 3s^2 3p^6 3d^6 4s^2$ (1)

Fe^{2+} (in $FeCl_2$) $1s^2 2s^2 2p^6 3s^2 3p^6 3d^6$ (1)

Fe^{3+} (in $FeCl_3$) $1s^2 2s^2 2p^6 3s^2 3p^6 3d^5$ (1)

> **Examiner's tip**
> When the atoms of d block elements in period 4 lose electrons to form ions, electrons are lost from the 4s sub-shell before the 3d sub-shell, even though the 3d sub-shell is filled after the 4s.

(d) Acidify the solution with dilute nitric acid. Then add silver nitrate solution. This will produce a white precipitate of silver chloride (1) which turns purple/grey in sunlight (1) /which dissolves in dilute ammonia solution.

> **Examiner's tip**
> Tests for Cl^- ions should always be acidified with dilute HNO_3 because OH^- and/or CO_3^{2-} ions in the solution would give a precipitate with $AgNO_3(aq)$.
> Dilute HNO_3 is chosen for the acidification because all nitrates are soluble and therefore NO_3^- ions will never produce a precipitate themselves.

(e) The high melting point of A (>500°C) suggests that it is <u>giant ionic with strong ionic bonds between Fe^{2+} and Cl^- ions.</u> (1) This is supported by the fact that A dissolves in water to form a <u>neutral solution containing $Fe^{2+}(aq)$ and $Cl^-(aq)$ ions.</u>   (1)

B has a relatively low melting point (220°C), which suggests a <u>simple molecular structure</u> (1) with a high degree of <u>covalent bonding</u>, this is supported by the <u>reaction with water to form an acidic solution.</u>   (1)

> **Examiner's tip**
> The acidic solution from **B** arises because the high charge density of Fe^{3+} causes water molecules attached to it as ligands to lose H^+ ions.
>
> $$[Fe(H_2O)_6]^{3+}(aq) \rightleftharpoons [Fe(H_2O)_5(OH)]^{2+}(aq) + H^+(aq)$$
>
> This process is sometimes called deprotonation and the equation can be written as
>
> $$[Fe(H_2O)_6]^{3+}(aq) + H_2O_{(1)} \rightleftharpoons [Fe(H_2O)_5OH]^{2+}(aq) + H_3O^+(aq)$$

Example B

(a) (i)   Scales, labels and units on axes   (1)

          Points plotted accurately   (1)

          Smooth curve through points   (1)

          End-point of titration $\cong$ 16.0 cm^3 (1) 0.050 mol dm^{-3} NaOH

> **Examiner's tip**
> The end-point is not at pH = 7, but at the mid-point of the rapidly rising part of the graph.

The graph suggests that A is monobasic. /1 mole of A reacts with 1 mole of NaOH.   *(1)*

$$\text{mass of A reacting} \quad = \frac{10}{1000} \times 7.2\,g = 0.072\,g \quad (1)$$

$$\text{Moles of NaOH reacting} \quad = \frac{16}{1000} \times 0.05 = 0.0008\,mol \quad (1)$$

$$\therefore\ 0.0008\,mol\ A \quad = 0.072\,g$$

$$\therefore \qquad M_r(A) \quad = \frac{0.072}{0.0008}\,g = 90\,g \quad (1)$$

(ii) Suppose the formula of A is HA.

$$HA(aq) \rightleftharpoons H^+(aq) + A^-(aq)$$

$$\therefore\ K_a = \frac{[H^+(aq)][A^-(aq)]}{[HA(aq)]} \quad \text{but}\quad [H^+(aq)] = [A^-(aq)]$$

$$\therefore\ K_a = \frac{[H^+(aq)]^2}{[HA(aq)]} \quad (1)$$

$$\text{pH of solution of HA} = 2.5 \ \therefore\ [H^+] = 3.16 \times 10^{-3}\,mol\,dm^{-3} \left.\right\} \ (1)$$

$$\text{Assuming HA is a weak acid } [HA(aq)] \cong \frac{7.2}{90} = 0.08\,mol\,dm^{-3}\left.\right\}$$

$$\therefore\ K_a = \frac{(3.16 \times 10^{-3})^2}{0.08} = 1.25 \times 10^{-4}\,mol\,dm^{-3} \quad (1)$$

(b) (i) Ratio of masses $C : H : O$ in A $= 40.0 : 6.70 : 53.3$

$$\text{Ratio of moles } C : H : O \text{ in A} \quad = \frac{40.0}{12} : \frac{6.70}{1} : \frac{53.3}{16}$$

$$= 3.33 : 6.70 : 3.33$$

$$= 1 : 2 : 1$$

$$\therefore\ \text{Empirical formula of A} = CH_2O \quad (1)$$

$$\text{Relative mass of empirical formula} = 30 \left.\right\} \ (1)$$

$$\therefore\ \text{Molecular Formula of A} = C_3H_6O_3 \left.\right\}$$

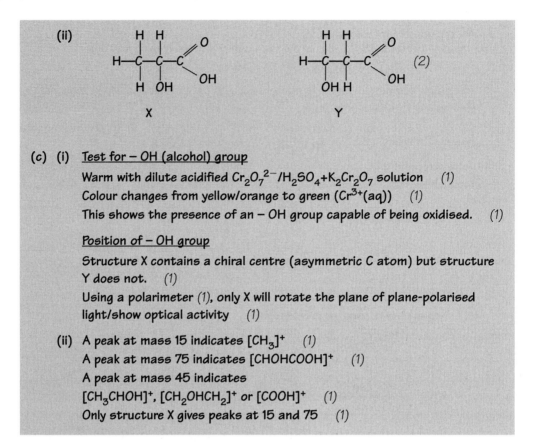

(c) (i) <u>Test for – OH (alcohol) group</u>

Warm with dilute acidified $Cr_2O_7^{2-}/H_2SO_4+K_2Cr_2O_7$ solution   *(1)*
Colour changes from yellow/orange to green ($Cr^{3+}(aq)$))   *(1)*
This shows the presence of an – OH group capable of being oxidised.   *(1)*

<u>Position of – OH group</u>

Structure X contains a chiral centre (asymmetric C atom) but structure Y does not.   *(1)*
Using a polarimeter *(1)*, only X will rotate the plane of plane-polarised light/show optical activity   *(1)*

(ii) A peak at mass 15 indicates $[CH_3]^+$ *(1)*
A peak at mass 75 indicates $[CHOHCOOH]^+$ *(1)*
A peak at mass 45 indicates
$[CH_3CHOH]^+$, $[CH_2OHCH_2]^+$ or $[COOH]^+$ *(1)*
Only structure X gives peaks at 15 and 75   *(1)*

Example C

(a) A spider diagram for Born–Haber cycles

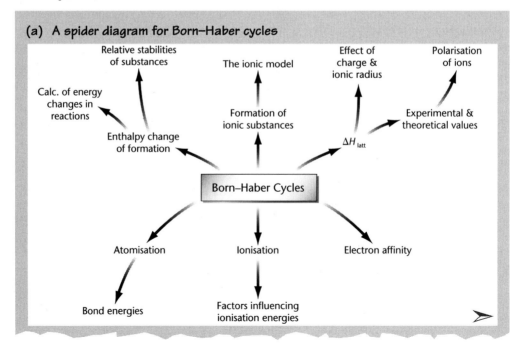

(b) A spider diagram for periodicity

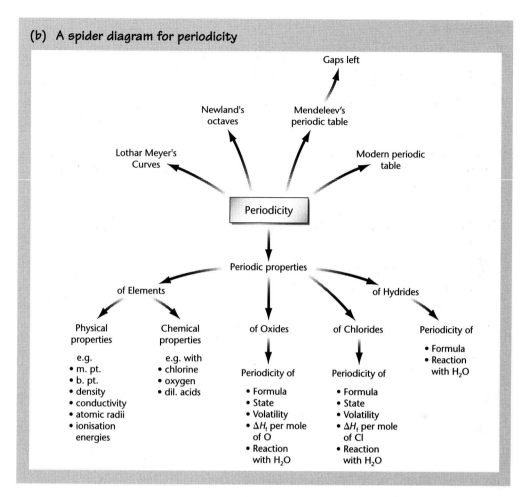

Example D

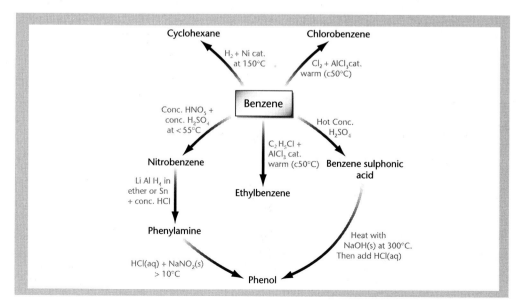

Example E

(a) $R-Br(l) + NaOH(aq) \xrightarrow{\text{heat}} R-OH(aq) + NaBr(aq)$ (1)

$NaBr(aq) + AgNO_3(aq) \longrightarrow AgBr(s) + NaNO_3(aq)$ (1)

Examiner's tip
State symbols, though not asked for in the question, have been included here because the introduction to the question makes the state of substances clear by using such words as 'aqueous' and 'precipitate'.

(b) **Weigh a small amount of the bromoalkane in a round bottom or pear-shaped flask.** *(1)*

Boil the bromoalkane under reflux *(1)* **using a vertically arranged Liebig's condenser** *(1)* **with excess sodium hydroxide solution.** *(1)*

Examiner's tip
Note that part (b) of the question asks for details of the apparatus, i.e. suitable flask, Liebig's condenser.

(c) **Allow the refluxed solution to cool.**
Then neutralise any excess NaOH(aq) <u>by adding dilute HNO₃</u> *(1)* **until the solution is acidic.** *(1)*
Now <u>add silver nitrate solution</u> *(1)* **to the mixture <u>until all the bromide is precipitated</u>** *(1)* **as silver bromide.**
Filter *(1)* **the mixture to collect the residue of AgBr.**
Wash *(1)* **the AgBr precipitate whilst it is in the filter paper.**
Then, allow the AgBr precipitate to dry *(1)* **on the filter paper in a warm room.**
Finally, weigh the AgBr precipitate. *(1)*   (max 7 marks)

Examiner's tip
Solids (precipitates) are normally removed from liquids and purified by filtering, washing and drying.

(d) **Suppose mass of dry AgBr is x g.**

Suppose mass of bromoalkane taken = y g

Mass of Br in the AgBr $= x \times \dfrac{A_r(Br)}{M_r(AgBr)}$ *(1)*

$\therefore$ **% Br in bromoalkane** $= \dfrac{x}{y} \times \dfrac{A_r(Br)}{M_r(AgBr)} \times 100$ *(1)*

Example F

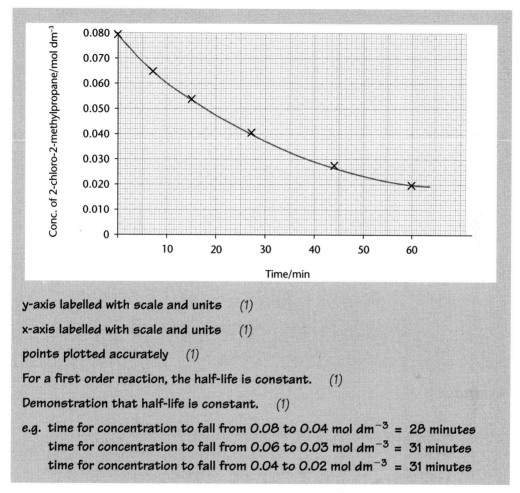

y-axis labelled with scale and units   *(1)*

x-axis labelled with scale and units   *(1)*

points plotted accurately   *(1)*

For a first order reaction, the half-life is constant.   *(1)*

Demonstration that half-life is constant.   *(1)*

e.g. time for concentration to fall from 0.08 to 0.04 mol dm^{-3} = 28 minutes
time for concentration to fall from 0.06 to 0.03 mol dm^{-3} = 31 minutes
time for concentration to fall from 0.04 to 0.02 mol dm^{-3} = 31 minutes

Example G

(a)

Oxide	Na_2O	MgO	Al_2O_3	SiO_2	P_4O_{10}	SO_3	
(i) Bonding	ionic	ionic	ionic	covalent	covalent	covalent	*(1)*
(ii) Structure	giant	giant	giant	giant	simple molecular	simple molecular	*(1)*

(iii) The melting point of MgO is very high (2827°C). In MgO, oppositely
charged ions (Mg^{2+} and O^{2-}) are held together by <u>strong ionic bonds</u>.   *(1)*

In comparison, the melting point of SO_3 is very low (33°C).
SO_3 consists of separate, discrete molecules of SO_3. Although the S and
three O atoms in each SO_3 molecule are held together by strong covalent
bonds, there are only <u>weak forces of attraction (van der Waals forces)</u>
between separate SO_3 molecules.   *(1)*

Thus SO_3 molecules can be separated easily and solid SO_3 melts at low temperatures, whereas the strong forces in MgO result in a high melting point. *(1)*

(b) (i) $SO_3 + H_2O \rightarrow H_2SO_4$ *(1)*

(ii) $Na_2O + H_2O \rightarrow 2NaOH$ *(1)*

Example H

Using a data book, $\Delta H_f^{\ominus}(SO_2(g)) = -297 \text{ kJ mol}^{-1}$ *(1)*

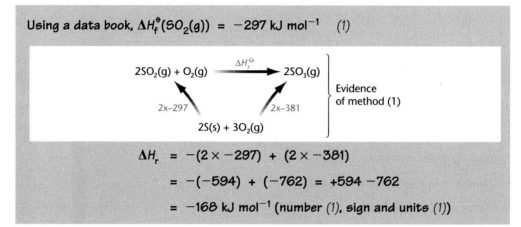

$$\Delta H_r = -(2 \times -297) + (2 \times -381)$$

$$= -(-594) + (-762) = +594 - 762$$

$$= -168 \text{ kJ mol}^{-1} \text{ (number (1), sign and units (1))}$$

Example I

(a) Ratio of masses of Co : N : H : Cl = 23.6 : 27.9 : 6.0 : 42.5

Ratio of moles of Co : N : H : Cl $= \dfrac{23.6}{58.9} : \dfrac{27.9}{14.0} : \dfrac{6.0}{1.0} : \dfrac{42.5}{35.5}$ *(1)*

$= 0.40 : 1.99 : 6.0 : 1.20$

Examiner's tip
Divide ratios of moles by the smallest value to get integral values.

Ratio of moles of Co : N : H : Cl $= \dfrac{0.40}{0.40} : \dfrac{1.99}{0.40} : \dfrac{6.0}{0.40} : \dfrac{1.20}{0.40}$

$= 1 : 5 : 15 : 3$ *(1)*

∴ Empirical formula of compound X $= CoN_5H_{15}Cl_3$ *(1)*

(b) (i) Air acts as an oxidising agent *(1)*, oxidising Co^{2+} to Co^{3+}. *(1)*

(ii) The ammonium chloride + ammonia acts as a buffer *(1)* and provides Cl^- ions. *(1)* (max 3 marks)

(c)   The electron shell structure of Co^{2+} in cobalt(II) chloride is 2,8,15.
This leaves space in the 3rd shell for co-ordinate bonding from ligands, such
as Cl^-, forming complex ions.   (1)
The electron shell structure of Ca^{2+} in calcium chloride is 2,8,8. This is a more
stable electron structure than that of Co^{2+} and co-ordinate bonding from Cl^-
does not occur.   (1)

<u>Other differences</u>

Co compounds can act as catalysts, unlike Ca compounds.   (1)
Co compounds occur in more than one oxidation state, whereas Ca
compounds occur in only the 2+ state.   (1)
Co^{2+} ions are coloured, whereas Ca^{2+} ions are white (or colourless in solution).   (1)
(max 4 marks)

> **Examiner's tip**
> The essential properties of transition metals, such as cobalt, which distinguishes them from
> other metals are:
> • catalytic activity of the elements and their compounds;
> • more than one oxidation state in their compounds;
> • coloured ions;
> • formation of complex ions.

Example J

(a) $E^{\ominus}/V$
Br$_2$(aq), 2Br$^-$(aq)/Pt   +1.09
Cl$_2$(aq), 2Cl$^-$(aq)/Pt   +1.36
I$_2$(aq), 2I$^-$(aq)/Pt   +0.54   (1)

(b)   The relevant electrode systems for the oxidation of iron by halogens are:

$Fe^{2+}(aq) \mid Fe(s)$ $E^{\ominus} = -0.44$ V

and $Fe^{3+}(aq), Fe^{2+}(aq)/Pt$ $E^{\ominus} = +0.77$ V   (1)

Chlorine is a strong oxidising agent. Its $E^{\ominus}$ value above indicates that it is
capable of oxidising Fe to Fe^{2+} and then Fe^{2+} to Fe^{3+}.   (1)

Iodine is also an oxidising agent. It is capable, like chlorine, of oxidising Fe to
Fe^{2+}.   (1)

$I_2(aq) + 2e^- \rightarrow 2I^-$ +0.54 V ⎤ overall
$Fe(s) \rightarrow Fe^{2+}(aq) + 2e-$ +0.44 V ⎦ $E^{\ominus} = +0.98$ V   (1)

Unlike chlorine, iodine is <u>not</u> capable of oxidising Fe^{2+} to Fe^{3+}.

$I_2(aq) + 2e^- \rightarrow 2I^-$ +0.54 V ⎤ overall
$Fe^{2+}(aq) \rightarrow Fe^{3+} + e^-$ −0.77 V ⎦ $E^{\ominus} = -0.23$ V   (1)

The negative $E^{\ominus}$ value indicates that the reaction is thermodynamically
unfeasible.   (max 5 marks)

Example K

(a) (i) Polarity associated with the alkene group and the <u>presence of an − OH group which can H-bond with water</u> *(1)* will make bombykol slightly soluble in water *(1)* in spite of the insolubility in water of the long hydrocarbon chain. *(1)* **(max 2 marks)**

   (ii) Geometric isomerism arises across each of the C=C bonds. So, there are four possible geometric isomers: *(1)* *cis–cis, cis–trans, trans–cis and trans–trans.* *(1)*

   (iii) Reactions:
   - with Na

$$CH_3(CH_2)_2CH=CHCH=CH(CH_2)_8CH_2OH + Na$$

$$\downarrow$$

$$CH_3(CH_2)_2CH=CHCH=CH(CH_2)_8CH_2O^-Na^+ \text{ (1)} + \tfrac{1}{2}H_2$$

   - with warm acidified $Cr_2O_7^{2-}/H_2SO_4 + K_2Cr_2O_7$

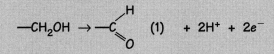 (1) $+ 2H^+ + 2e^-$

   Then 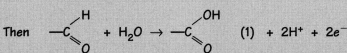 (1) $+ 2H^+ + 2e^-$

   - with conc. HCl

$$-CH_2OH + H^+ + Cl^- \rightarrow -CH_2Cl \text{ (1)} + H_2O$$

   - with excess conc. H_2SO_4 at 170°C

$$-CH_2-CH_2OH \rightarrow -CH=CH_2 \text{ (1)} + H_2O$$

   - with CH_3COOH + conc. H_2SO_4 catalyst

 (1) $+ H_2O$

   (1 mark for each correct organic product: max 4 marks)

Examiner's tip
In part (iii) above, only the possible reactions of the primary −OH group in bombykol have been shown. There is another series of possible products from reactions of the two alkene (C=C) groups.

(b) $CH_3(CH_2)_2COOH$,   butanoic acid
   $COOHCOOH$,   ethanedioic acid
   $HOOC(CH_2)_8COOH$, decanedioic acid
   Names or formulae of 3 acids   *(2)*
   Names or formulae of 2 acids   *(1)*

Example L

Ammonia

Structure and bonding

Electronic structure of N is 2,5 $\Big\}$ *(1)*
Electronic structure of H is 1

∴ N forms three covalent bonds with three H atoms to form NH_3. *(1)* (By doing this, the N and H atoms gain electron structures like noble gases.)

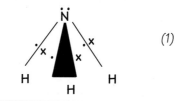

(1)

The N atom has three covalent bonds and one lone pair in its outer shell. Hence its structure is pyramidal. *(1)*

Lone pairs of electrons exert a greater repelling force than shared pairs (in covalent bonds) so the H–N–H bond angle is less than the normal tetrahedral angle of $109\frac{1}{2}°$. In NH_3 the H–N–H bond angle is 105°. *(1)* (max 4 marks)

Manufacture

From nitrogen (obtained by fractional distillation of liquid air *(1)*) and hydrogen (obtained from natural gas or naphtha *(1)*)

$$N_2(g) + 3H_2(g) \rightarrow 2NH_3(g) \quad (1)$$

The reaction uses an Fe catalyst *(1)* at 250 atm *(1)* and 450°C. *(1)*

The reaction results in an equilibrium favoured by high pressure and low temperature. *(1)*

Low temperatures would result in a reaction rate that is too slow and so a compromise temperature of 450°C is employed. *(1)* (max 5 marks)

Reactions – inorganic

- Reactions as a base with acids *(1)*
  e.g. $NH_3 + HCl \rightarrow NH_4^+Cl^-$ *(1)*

- Reactions as a weak alkali, precipitating metal hydroxides *(1)*
  ($NH_3 + H_2O \rightleftharpoons NH_4^+ + OH^-$)
  e.g. $Cu^{2+}(aq) + 2OH^-(aq) \rightarrow Cu(OH)_2(s)$ *(1)*

- Reactions as a ligand, forming complexes with metal ions *(1)*
  e.g. $Cu^{2+}(aq) + 4NH_3(aq) \rightarrow [Cu(NH_3)_4]^{2+}(aq)$ *(1)*

- Formation of co-ordinate bonds
  e.g. with H^+, $AlCl_3$, etc. *(1)*
  $NH_3 + H^+ \rightarrow NH_4^+$
  or $NH_3 + AlCl_3 \rightarrow H_3N{:}{\rightarrow}AlCl_3$ *(1)*

➤

<u>Reactions – organic</u>

- Reactions as a base with organic acids   *(1)*

  $NH_3 + CH_3COOH \rightarrow CH_3COO^- NH_4^+$ *(1)*

- Reactions as a nucleophile with halogenoalkanes   *(1)*

  $RCl + NH_3 \rightarrow RNH_2 + HCl \rightarrow RNH_3^+Cl^-$ *(1)*

                           amine

- Reactions as a nucleophile with acyl halides   *(1)*

  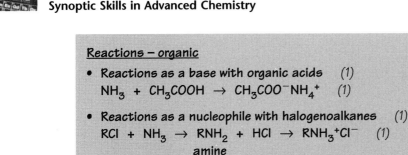

                        amide

(The HCl produced will react with excess NH_3 to form NH_4Cl).

(max 6 marks)

> **Examiner's tip**
>
> *In marking essays, there are various possible approaches and the mark scheme must take these into account. In the last part of Example L, there are in fact 14 marks shown on the mark scheme.*
>
> *Notice, however, that three clear statements of the type of reaction with a related equation will gain the maximum 6 marks.*

Mark schemes for the mock examination papers

Although I have seen the specimen mark schemes prepared by the awarding bodies for some of the questions in the following mock examination papers, the mark schemes in this book are entirely my responsibility

Paper 1 (Unit Test 1) Mark Scheme and Examiner's Tips

Multiple choice questions

1   Correct answer is **B**.

> **Examiner's tip**
>
> *The maximum oxidation number of an element is usually equal to the number of electrons beyond that of the previous noble gas.*
>
> *So, for example, A has 5 electrons beyond He (electron structure $1s^2$) so A has a maximum oxidation number of +5. A is nitrogen.*

2   Correct answer is **D**.

> **Examiner's tip**
>
> *Molecules and ions with zero dipole moment are symmetrical about a central point in their structure.*
>
> *NH_4^+ has four N–H bonds arranged tetrahedrally.*

3 Correct answer is **C**.

From the half-equations:

5 mol of H_2O_2 ≡ 2 mol of $KMnO_4$

∴ 0.002 mol of H_2O_2 ≡ $\dfrac{2}{5}$ × 0.002 mol $KMnO_4$ = $\dfrac{v}{1000}$ × 0.02 mol $KMnO_4$

where v is the volume in cm^3 of 0.02 mol dm^{-3} $KMnO_4$

$\Rightarrow \dfrac{v}{1000} \times 0.02 = \dfrac{2}{5} \times 0.002$

∴ $v = \dfrac{2}{5} \times \dfrac{0.002}{0.02} \times 1000 = 40\ cm^3$

> *Examiner's tip*
> *In order to balance half-equations, the number of electrons given and taken by the reductant and oxidant must balance. So, $5H_2O_2$ will give up 10 electrons and these would be taken by $2MnO_4^-$.*

4 Correct answer is **D**.

> *Examiner's tip*
> *Entropy is roughly equivalent to disorder. So, a decrease in entropy is a decrease in disorder or an increase in order, and involves gases → liquids, liquids → solids, fewer molecules being produced or weakly bonded substances → strongly bonded substances.*
>
> *D involves a decrease in entropy with 4 mols forming 2 mols and a mixture of two substances (N_2 and H_2) forming just one substance (NH_3).*

5 Correct answer is **A**.

> *Examiner's tip*
> *Across a period, electrons are going into the same outer shell whilst the positive charge in the nucleus increases. The increasing nuclear charge attracts the electrons more strongly, causing a decrease in the atomic radius.*

6 Correct answer is **B**.

> *Examiner's tip*
> *Increasing the temperature will increase the rate of the reaction. As concentrations have not changed, this means that there must be an increase in the rate constant.*

7 Correct answer is **B**.

BF_3 is trigonal planar, NH_3 is pyramidal.
$BeCl_2$ and HCN are both linear.
H_2O is V shaped.
SO_2 is V shaped and CO_2 is linear.

8 Correct answer is **C.**

> *Examiner's tip*
> *There are two kinds of stereoisomerism: geometric (cis/trans) isomerism and optical isomerism.*
>
> *Geometric isomers have a double bond in which the two carbon atoms across the double bond must both have two different groups attached to them.*
>
> *Optical isomers must have a chiral centre (this commonly involves having a carbon atom with four different groups attached to it).*
>
> *This is true of the second carbon atom in C.*

9 Correct answer is **D.**

> *Examiner's tip*
> *Both aldehydes and alcohols can be oxidised by dilute acidified potassium dichromate(VI) and dilute acidified potassium manganate(VII).*
>
> *Ketones, such as CH_3COCH_3, are not oxidised by acidified potassium dichromate(VI) or dilute acidified potassium manganate(VII).*

10 Correct answer is **A.**

The oxidation numbers of carbon and chlorine change in alternative A, showing that it involves redox.

> *Examiner's tip*
> *Always use oxidation numbers to check for redox.*

Multiple completion questions

11 Correct answer is **D.**

> *Examiner's tip*
> *Warm dilute acidified $KMnO_4$ is a good oxidising agent.*
>
> *It will, for example, oxidise alkenes to diols, alcohols to aldehydes and ketones, aldehydes to carboxylic acids, Fe^{2+} to Fe^{3+}, Sn^{2+} to Sn^{4+} and I^- to I_2. It cannot oxidise Fe^{3+} or Cu^{2+} as these are the highest stable oxidation states of iron and copper, respectively. Nor can it oxidise ketones.*

12 Correct answer is **C.**

> *Examiner's tip*
> *Hydrogen bonding occurs in compounds in which a hydrogen atom is bonded directly to one of the three most electronegative elements (N, O and F).*
>
> *In CH_3CHO and CH_2F_2, hydrogen atoms are not attached to N, O or F. The displayed formulas of these two compounds are:*

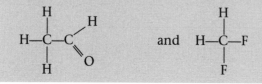

13 Correct answer is **A**.

Acids are proton (H^+ ion) donors.
Bases are proton acceptors.
In this reaction, one water molecule acts as an acid and the other acts as a base.

14 Correct answer is **C**.

CH_3Cl, like other chloroalkanes, is insoluble in water.
CH_3COCl, like other acyl chlorides, reacts with water to form a carboxylic acid and hydrogen chloride (which dissolves in the water to form hydrochloric acid).

$$CH_3COCl + H_2O \rightarrow CH_3COOH + HCl$$

PH_3 is weakly alkaline. (Both P and N are in group 5 and PH_3 is similar to NH_3.)
PCl_3 reacts with water to from phosphorous acid, H_3PO_3, and hydrogen chloride. The hydrogen chloride reacts with excess water to form hydrochloric acid.

$$PCl_3 + 3H_2O \rightarrow H_3PO_3 + 3HCl$$

15 Correct answer is **A**.

Displayed formulae for the four substances are:

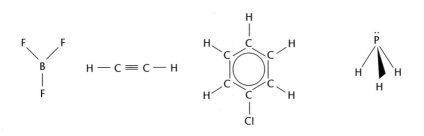

Examiner's tip
Draw displayed formulae in order to check the structure of simple molecules. Show lone pairs on atoms attached to more than one other atom (e.g. only P in the formulae above).

Atoms with only two negative centres around them (e.g. the two C atoms in H–C≡C–H which have a single covalent bond and a triple covalent bond) will have a linear arrangement of the negative centres.

Atoms with three negative centres around them (e.g. B in BF_3 and all six C atoms in ⬡–Cl) will have a trigonal planar arrangement of the negative centres.

Atoms with four negative centres around them (e.g. P in PH_3 with three single covalent bonds and one lone pair) will have a tetrahedral arrangement of the negative centres although, in this case, a pyramidal arrangement of atoms.

16 Correct answer is **A**.

The elements close to chlorine in the periodic table with their electron structures are:

	Al	Si	P	S	Cl	Ar
	2,8,3	2,8,4	2,8,5	2,8,6	2,8,7	2,8,8

	K	Ca
	2,8,8,1	2,8,8,2

The electronic structures of Cl^-, Ca^{2+} and S^{2-} are all 2,8,8 with 18 electrons. The same applies to PH_3.

> **Examiner's tip**
> When an element forms one covalent bond to hydrogen, the product has one more electron and, of course, one more proton.

17 Correct answer is **B**.

> **Examiner's tip**
> In every group of the periodic table, elements become more electropositive (more metallic or less non-metallic) and have smaller, less endothermic first ionisation energies with increasing relative atomic mass.

18 Correct answer is **B**.

From experiments 1, 2 and 3, the reaction is second order with respect to A. From experiments 3, 4 and 5, the reaction is zero order with respect to B.

> **Examiner's tip**
> If a reaction is zero order with respect to one reactant, the reaction must occur in at least two steps with at least one step, involving the zero order reactant, which is not the rate-determining (slowest) step for the reaction.

19 Correct answer is **C**.

Both ethanol and phenol form salts with sodium (sodium ethoxide and sodium phenoxide, respectively) and esters with ethanoyl chloride (ethyl ethanoate and phenyl ethanoate).
Phenol reacts with bromine water (forming 2,4,6–tribromophenol) and with neutral $FeCl_3(aq)$ to form a violet coloured complex ion. Ethanol does <u>not</u> react with bromine water or with $FeCl_3(aq)$.

20 Correct answer is **D**.

Conc. H_2SO_4 will dehydrate ethanol forming ethene, but it has no reaction with ethanoic acid.
Potassium manganate(VII) can oxidise ethanol to ethanal and then to ethanoic acid but it has no reaction with ethanoic acid.
Sodium hydroxide has no reaction with ethanol but it forms sodium ethanoate with ethanoic acid.
Sodium reacts with ethanol to form sodium ethoxide plus <u>hydrogen</u> and with ethanoic acid to form sodium ethanoate plus <u>hydrogen</u>.

Paper 2 (Unit Test 2) Mark Scheme and Examiner's Tips

1 (a) (i) $pH = -\log_{10}[H^+]$ or $pH = -\lg [H^+]$ (1)
$[H^+]$ may be replaced by $[H_3O^+]$

 (ii) $K_a = \dfrac{[H^+(aq)]_{eqm} [HCO_3^-(aq)]_{eqm}}{[CO_2(aq)]_{eqm}}$ correct numerators and denominator (1)
square brackets and state symbols (1)

> *Examiner's tip*
> *Some examiners will expect you to show the subscript 'eqm' after each concentration term in the expression for K_a, although the normally accepted practice is to omit them.*

 (b) (i) From the expression for K_a
$4.5 \times 10^{-7} = \dfrac{[H^+(aq)] \, 2.5 \times 10^{-2}}{1.25 \times 10^{-3}}$ evidence of calculation (1)

$\Rightarrow [H^+(aq)] = 4.5 \times 10^{-7} \times \dfrac{1.25 \times 10^{-3}}{2.5 \times 10^{-2}} = \underline{2.25 \times 10^{-8} \text{ mol dm}^{-3}}$ (1)

 (ii) $\therefore$ pH of blood of a healthy person = 7.65 (1)

 (c) If a small amount of acid gets into the blood, <u>the extra H^+ ions react with HCO_3^- ions in the blood (1)</u>, the equilibrium in **equation 1** moves to the <u>left (1)</u> and $[H^+]$ does not change significantly.
If a small amount of alkali (i.e. OH^- ions) get into the blood, <u>the extra OH^- ions react with H^+ ions in the blood to form water.</u> (1)

$$H^+(aq) + OH^-(aq) \rightarrow H_2O(l) \quad (1)$$

Equation 1 then moves to the right to restore the equilibrium (1) and $[H^+]$ does not change significantly.

 (d) $H^+(aq) + HCO_3^-(aq) \rightleftharpoons 2H^+(aq) + CO_3^{2-}(aq)$ (1)
(or $HCO_3^-(aq) \rightleftharpoons H^+(aq) + CO_3^{2-}(aq)$)
$Ca^{2+}(aq) + CO_3^{2-}(aq) \rightleftharpoons CaCO_3(s)$ (1)

> *Examiner's tip*
> *Strictly, the equations above should have equilibrium arrows as shown, but straightforward arrows would probably not be penalised.*

2 (a) (i) When $[H_2]$ increases 4 times, the rate increases 4 times.
$\therefore$ order with respect to H_2 is 1. (1)
When $[I_2]$ increases 3 times, the rate increases 3 times.
$\therefore$ order with respect to I_2 is 1. (1)
Statement of relationship of rate increase resulting from concentration increase for either H_2 or I_2 (1)

 (ii) Rate $= k [H_2][I_2]$ (1)

(iii) The <u>rate is dictated by the slowest step in the reaction mechanism</u> (1) whereas the <u>stoichiometric equation is the sum of all the steps in the mechanism.</u> (1)

(or the rate equation involves only those particles which affect the reaction rate (1) whereas the stoichiometric equation involves all particles which take part in the reaction. (1))

(iv) The rate-determining step is the <u>slowest step in the mechanism for the reaction.</u> (1) (or the step which dictates the overall reaction rate.)

(v) For the reaction $A_2 + B_2 \rightarrow 2AB$
suppose rate $= k[A_2]$ i.e. first order with respect to A_2; zero order with respect to B_2 (1) Mechanism

$$A_2 \xrightarrow{\text{slow}} A + A \quad (1)$$
$$B_2 \xrightarrow{\text{fast}} B + B \quad (1)$$
$$\text{then } A + B \xrightarrow{\text{fast}} AB$$

> **Examiner's tip**
> If a reaction is zero order with respect to a particular reactant, that reactant (or products from it) cannot be involved in the rate-determining step.

(b) (i) The reactant particles are moving faster and therefore <u>collide more frequently.</u> (1)
The reactant particles <u>collide with more energy</u> (1) and therefore more collisions result in reaction.

(ii)

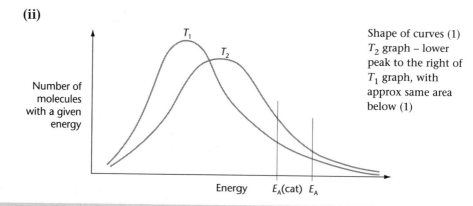

Shape of curves (1)
T_2 graph – lower peak to the right of T_1 graph, with approx same area below (1)

> **Examiner's tip**
> Note the general shapes of the two curves at temperatures. T_1 and T_2 above $(T_2 > T_1)$ and the relative positions of E_A and $E_A(cat)$. You are now in a good position to explain the effects of temperature and catalysts on reaction rate.

E_A represents activation energy, i.e. <u>the energy needed by colliding particles for a reaction</u> to occur. (1)
At a higher temperature (T_2), the fraction of molecules with energy sufficient to react (i.e. energy $\geq E_A$) is greater. (1)
$\therefore$ the reaction rate is greater.

(iii) <u>When a catalyst is used, the activation energy is lower</u> (i.e. E_A(cat)).
(1) There are <u>now more molecules with energy $\geq E_A$(cat)</u> (1) and thus the reaction rate increases.

3 (a)

$Na^+(g) + e^- + Br(g)$

$\frac{1}{2}\Delta H^{\ominus}_{diss}(Br_2(g))$ $\Delta H^{\ominus}_{ea}$

$Na^+(g) + e^- + \frac{1}{2}Br_2(g)$ $Na^+(g) + Br^-(g)$

$\frac{1}{2}\Delta H^{\ominus}_{vap}(Br_2(l))$

$Na^+(g) + e^- + \frac{1}{2}Br_2(l)$

$\Delta H^{\ominus}_i$

$Na(g) + \frac{1}{2}Br_2(l)$ $\Delta H^{\ominus}_L$

$\Delta H^{\ominus}_{sub}$

$Na(s) + \frac{1}{2}Br_2(l)$

$\Delta H^{\ominus}_f$

$Na^+ + Br^-(s)$

(1) for each correctly labelled step in the cycle, up to a maximum of 6 marks.

Examiner's tip

Notice that the data provided gives the lattice dissociation of NaBr(s) not the lattice energy. The lattice dissociation is the reverse of the lattice energy.

$$Na^+Br^-(s) \xrightarrow[\text{lattice energy}]{\text{lattice dissociation}} Na^+(g) + Br^-(g)$$

In addition, the process which the question describes as 'electron addition enthalpy' is usually called 'electron affinity'.

*Notice also that $\Delta H^{\ominus}_{diss}$ relates to the bond dissociation of $Br_2(g)$
i.e. to the reaction $Br_2(g) \rightarrow 2Br(g)$
The process $\frac{1}{2}Br_2(g) \rightarrow Br(g)$ is therefore $\frac{1}{2}\Delta H^{\ominus}_{diss}$.*

*Likewise, the enthalpy of vaporisation of liquid bromine, $\Delta H^{\ominus}_{vap}$ will relate to the vaporisation of one mole of $Br_2(l)$.
i.e. $Br_2(l) \rightarrow Br_2(g)$*

Born–Haber cycles can be simplified by first considering just three steps, as in the energy cycle below.

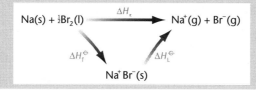

ΔH_x *in this energy cycle can be broken down into two separate stages:*

$$Na(s) \rightarrow Na^+(g) + e^- \text{ and}$$
$$\tfrac{1}{2}Br_2(l) + e^- \rightarrow Br^-(g)$$

The stage involving Na(s) is completed in two steps, shown at the lower left of the full cycle. The stage involving $\tfrac{1}{2}Br_2(l)$ follows in the next three steps of the full cycle.

These three steps from $\tfrac{1}{2}Br_2(l)$ to $Br^-(g)$ are sometimes reduced to just two steps in Born–Haber cycles.

i.e. $\tfrac{1}{2}Br_2(l) \xrightarrow{\Delta H^{\ominus}_{at}(\tfrac{1}{2}Br_2(l))} Br(g) \xrightarrow{\Delta H^{\ominus}_{ea}(Br(g))} Br^-(g)$

$\Delta H^{\ominus}_{at}(\tfrac{1}{2}Br_2(l))$ *is the standard enthalpy change of atomisation of bromine.*

(b) $+107 + 498 + \tfrac{1}{2}\Delta H^{\ominus}_{vap}(Br_2(l)) + \tfrac{1}{2} \times 194 - 325 = -361 + 753$
(one equation correctly linking the values (2); up to 2 errors in an equation linking values (1))

$$\therefore \tfrac{1}{2}\Delta H^{\ominus}_{vap}(Br_2(l)) = -361 + 753 - 107 - 498 - 97 + 325$$

$$= +15 \text{ kJ mol}^{-1}$$

$$\therefore \Delta H^{\ominus}_{vap}(Br_2(l)) = +30 \text{ kJ mol}^{-1} \quad (1)$$

4 (a) Mass spectroscopy

X should produce peaks due to $COOH^+$ (rel. mass 45), $C_6H_4CH_2NH_2^+$ (rel. mass 106), $C_6H_4COOH^+$ (rel. mass 121), $CH_2NH_2^+$ (rel. mass 30), $C_6H_4^+$ (rel. mass 76), etc. (1)
Y should produce peaks due to $COOH^+$ (rel. mass 45), $C_6H_5CHNH_2^+$ (rel. mass 106), $C_6H_5^+$ (rel. mass 77), $CHNH_2COOH$ (rel. mass 74), etc. (1)

Peaks at 30 and 76 will identify X (1)
Peaks at 74 and 77 will identify Y (1)   (max 3 marks)

Infra-red spectroscopy

IR spectroscopy will identify different types of bonds. (1)
e.g. Using the Nuffield Advanced Science, Book of Data,
N–H absorbs at wave number 3500–3300 cm^{-1}
O–H in carboxylic acids absorbs at 3300–2500 cm^{-1}
C–H in arenes absorbs at 3030 cm^{-1}, etc. at least 2 pieces of data for (1)

X and Y contain exactly the same types of bonds and so IR spectroscopy cannot distinguish between X and Y. (1)   (max 3 marks)

NMR spectroscopy

NMR will identify H atoms in different environments and <u>show the ratio of the number of H atoms in different environments.</u> (1)

Environments	N–H, alkyl	C–H,	arene C–H,	carboxylic acid O–H	
Ratio for **X**	2	2	4	1	(1)
Ratio for **Y**	2	1	5	1	(1) (max 3 marks)

(Overall a maximum of 3 marks for any one method and an overall maximum for section **(a)** of 6.)

Examiner's tip
Notice how the three spectroscopic methods are well supported with data as the question requires.

(b) **Y** is chiral.   (1)

It therefore has optical isomers which are optically active, <u>rotating the plane of plane-polarised (monochromatic) light</u> (1) in opposite directions.

(c)

$$\left(\begin{array}{c} \overset{H}{\underset{|}{N}}-CH_2-\text{⬡}-\overset{O}{\overset{\|}{C}}-\overset{H}{\underset{|}{N}}-CH_2-\text{⬡}-\overset{O}{\overset{\|}{C}}- \end{array} \right)_n \quad (1)$$

The other product of the process is HCl.   (1)

Paper 3 (Unit Test 3) Mark Scheme and Examiner's Tips

1 **(a)** **(i)**   Le Chatelier's principle says that a system in equilibrium will respond to a constraint (change) by trying to reduce or remove that constraint.   (1)
So, in this case, the optimum conditions are:

- <u>low temperature</u> (1) because low temperatures will move the system towards the product, <u>evolving heat</u> (the reaction is exothermic) <u>to try to restore the temperature</u>   (1)
- <u>high pressure</u> (1) because increased pressure will move the system towards the product as <u>2 moles of reactant gases react to form just 1 mole of product gas so reducing the pressure.</u> (1) (max 4 marks)

Examiner's tip
In predicting the optimum conditions for the formation of the product of a system in equilibrium, you are expected to use Le Chatelier's principle to suggest the best conditions of <u>temperature and pressure</u>. In this case, you should also justify your predictions by referring to and applying Le Chatelier's principle.

(ii)   Low temperatures may reduce the reaction rate to a level which is too slow.   (1)
∴ the actual conditions may have to use a compromise temperature.   (1)
High pressure may require <u>equipment which is too expensive or conditions which are unsafe</u> (1) so a lower compromise pressure may be used.   (max 2 marks)

(iii) The system can be cooled to say 50°C.
<u>Water and ethanol will condense</u> as a mixture of liquids (1) from which <u>ethanol can be obtained by fractional distillation</u>. (1)

(b) (i) $K_p = \dfrac{(P_{C_2H_5OH})_{eqm}}{(P_{C_2H_4})_{eqm} \times (P_{H_2O})_{eqm}}$ (1)

    (ii) Increase in temperature <u>will reduce K_p</u> (1) because the equilibrium will move to the left. (1)
Increase in pressure will push the equilibrium to the right, but <u>this will change the pressures of all the reactants and products</u> (1) and the value of K_p will <u>not</u> change.
The presence of a catalyst will speed up the forward and reverse reactions equally (1) but the value of K_p will <u>not</u> change.
(Allow 1 mark if you have said that K_p is unaffected by both change in pressure and presence of a catalyst without explanation.)

(c) (i) Using octane as the typical alkane in petrol:
$C_8H_{18}(g) + 12\tfrac{1}{2}O_2(g) \rightarrow 8CO_2(g) + 9H_2O(g)$ (1)
$C_2H_5OH(g) + 3O_2(g) \rightarrow 2CO_2(g) + 3H_2O$ (1)

    $\Rightarrow$ 114 g octane requires 400 g O_2 $\therefore$ 1 g octane requires 3.51 g O_2 (1)
        46 g ethanol requires 96 g O_2 $\therefore$ 1 g ethanol requires 2.09 g O_2 (1)

    (ii) Both fuels produce CO_2 and H_2O on complete combustion. Ethanol, unlike octane, <u>contains some oxygen in its own molecules</u> (1) so it requires less oxygen from the air per gram to form CO_2 and H_2O.

2 (a) (i) Electronic configuration Cu $1s^2 2s^2 2p^6 3s^2 3p^6 3d^{10} 4s^1$ (1)
Electronic configuration Cu$^+$ $1s^2 2s^2 2p^6 3s^2 3p^6 3d^{10}$ (1)

    (ii) Across the first transition series, elements tend to increase the number of electrons in the 3d sub-shell whilst retaining two electrons in the 4s sub-shell.
<u>Copper might be expected to have the structure [Ar]3d^{9}4s^2</u> (1) rather than [Ar]3d^{10}4s^1.

Examiner's tip
Electronic configurations can be abbreviated by writing the symbol of the previous noble gas in square brackets.

(iii) Colour in transition metal ions is due to a partially filled d sub-shell (1) within which electron transitions can take place. Cu^+ has a full d sub-shell.

(b) If Cu_2SO_4 is added to water, disproportionation (1) would occur. Cu^+ ions will, at the same time, be oxidised to Cu^{2+} and reduced to Cu. (1)

$$
\begin{array}{lll}
\text{i.e.} & Cu^+(aq) \rightarrow Cu^{2+}(aq) + e^- & \left.\begin{array}{l} -0.15 \\ +0.52 \end{array}\right\} \begin{array}{l} \text{overall} \\ E = +0.37 \text{ V} \quad (1) \end{array} \\
& \underline{Cu^+(aq) + e^- \rightarrow Cu(s)} & \\
& 2Cu^+(aq) \rightarrow Cu^{2+}(aq) + Cu(s) &
\end{array}
$$

E/V above the bracket.

(2 half-equations or the overall equation (1))

Examiner's tip
E and $E^{\ominus}$ values relate to the direction in which the equation is written. If the reaction/equation is reversed, then the E or $E^{\ominus}$ value changes its sign.

When the reaction occurs, the solution would become blue (1) (due to $Cu^{2+}(aq)$ ions) and a solid precipitate of red/brown copper (1) will form.

(c) (i)

	3d	4s	4p	4d
[Ar]	↑↓ ↑↓ ↑↓ ↑↓ ↑	↑↓	↑↓ ↑↓ ↑↓	↑↓ ↑↓ □ □

$\underbrace{\qquad}_{Cu^{2+}\text{ electrons (1)}}$ $\underbrace{\qquad}_{\text{Ligand electrons (1)}}$

Examiner's tip
In a complex ion, each ligand donates one pair of electrons into orbitals of the cation.

So, in this case, the Cu^{2+} ion receives six electron pairs, one pair from each of the H_2O molecules.

(ii) A pale blue precipitate (1) will form with a little $NH_3(aq)$.

$$[Cu(H_2O)_6]^{2+}(aq) + 2OH^-(aq) \rightarrow [Cu(H_2O)_4(OH)_2](s) + 2H_2O(l) \quad (1)$$
$$\text{pale blue ppte}$$

This is a precipitation reaction/acid–base reaction. (1)

With excess $NH_3(aq)$, a deep blue solution (1) forms.

$$[Cu(H_2O)_4(OH)_2](s) + 4NH_3(aq) \rightarrow [Cu(H_2O)_2(NH_3)_4]^{2+}(aq) + 2H_2O(l) + 2OH^-(aq) \quad (1)$$

This is a ligand exchange reaction/reaction involving complexes. (1)

(max 5 marks)

> *Examiner's tip*
>
> *The aqueous Cu^{2+} ion is represented most accurately as either $[Cu(H_2O)_6]^{2+}(aq)$ or $[Cu(H_2O)_4]^{2+}(aq)$. Hence the equations used.*
>
> *If you have represented the aqueous Cu^{2+} ion as $Cu^{2+}(aq)$ and written the equations as:*
>
> $Cu^{2+}(aq) + 2OH^-(aq) \rightarrow Cu(OH)_2(s)$ *and*
>   *pale blue ppte*
>
> $Cu(OH)_2(s) + 4NH_3(aq) \rightarrow [Cu(NH_3)_4]^{2+}(aq) + 2OH^-(aq)$
>
> *you would probably get full credit for the first equation showing precipitation of $Cu(OH)_2$, but not for the second.*

3 (a) (i) Pass Cl_2 gas (1) into warm (1) benzene in the presence of a catalyst of $AlCl_3$ (1) or $FeCl_3$ or Fe (which reacts first to form $FeCl_3$).

    (ii) The catalyst <u>polarises the Cl_2</u> molecules. (1)

                    δ+   δ−   δ+   δ−

          i.e. Cl—Cl·····$Al^{3+}(Cl^-)_3$ / Cl—Cl·····$Fe^{3+}(Cl^-)_3$ (1)

          due to the <u>high positive charge density on the Al^{3+}/Fe^{3+} ion</u>. (1)
This enables the <u>chlorine atom that is polarised δ+ to attack the benzene ring</u>. (1)   (max 3 marks)

(b) Shake the water sample in a separating funnel (1) with ethyl ethanoate. (1)
Allow the liquids to separate, run off the lower layer of water. (1)
Then shake the ethyl ethanoate with a further volume of water.
(The process can be repeated with further samples of water.) (1)
Finally, run off the lower layer of water and then run off the ethyl ethanoate plus washings into a round bottom flask.
Heat the ethyl ethanoate containing herbicides carefully <u>to evaporate the solvent and concentrate the herbicides</u>. (1)

    (i) The methyl ester will be more volatile than the carboxylic acid. (1)
The methyl ester is therefore less likely to condense in the chromatography column and will move through the column faster. (1)

    (ii)

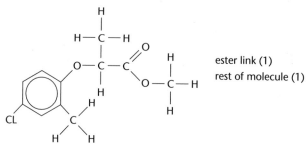

ester link (1)
rest of molecule (1)

Paper 4 (Unit Test 4) Mark Scheme and Examiner's Tips

1 **(a)** No. of moles of H_2O_2 decomposed $= \dfrac{100}{1000} \times 1.00 = 0.10$ (1)

∴ No. of moles of oxygen produced $= 0.050$ (1)
Using the ideal gas equation, $pV = nRT$ (1)
$100\,000 \times V = 0.05 \times 8.31 \times 293$ (1)
$\Rightarrow V = 1.22 \times 10^{-3}$ m^3/1.22 dm^3 (1)

> ***Examiner's tip***
> *As the value of the gas constant has been given, the examiner would expect you to use the ideal gas equation to calculate the volume of oxygen.*
>
> *The gas constant is, of course, given in SI units and so all the quantities substituted in the equation should be in SI units.*
>
> *So, 100 kPa = 100 000 Pa = 100 000 N m^{-2} and*
>
> 20°C = 293 K
>
> *The equation will then give a value for the volume in m^3 (i.e. SI units).*
>
> *As an alternative, but less accurate, method of finding the volume of oxygen, it is possible to use the fact that 1 mole of gas at 20°C and 100 kPa (1 atm) will occupy approx 24 dm^3.*
>
> $\Rightarrow$ *0.05 moles of O$_2$ will occupy 0.05 × 24 = 1.2 dm^3*
>
> *This approach would gain a maximum of 4 marks.*

(b) <u>Apparatus</u>

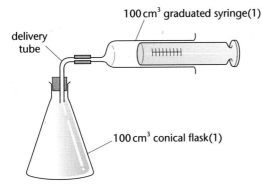

100 cm^3 graduated syringe(1)

delivery tube

100 cm^3 conical flask(1)

<u>Amount of H_2O_2 solution to use:</u>
100 cm^3 of the 1.0 mol dm^{-3} H_2O_2 solution gives 1.22 dm^3 (1220 cm^3). As the syringe holds only 100 cm^3 (1), it would be sensible to dilute the H_2O_2 to 0.10 mol dm^{-3} and then use between 50 and 80 cm^3 of this. (1)

Procedure:

- weigh out the same number of moles (say 0.01 moles) of each oxide (1)
- add 50 cm³ of 0.01 mol dm⁻³ H_2O_2 solution rapidly (1) to 0.01 moles of oxide I, start the stop watch (1), quickly replace the bung carrying the delivery tube and mix the contents of the flask thoroughly (1)
- measure the volume of O_2 produced in the first two minutes (1) after adding the H_2O_2 solution.
- repeat the experiment in exactly the same manner and conditions (1), replacing oxide I with oxide II (1).

(max 10 marks)

> **Examiner's tip**
>
> Notice that the mark scheme above indicates that it is sensible to dilute the 1.00 mol dm⁻³ H_2O_2 in section **(a)** for the experiment in section **(b)**.
>
> If you chose to use the 1.00 mol dm⁻³ H_2O_2, the maximum volume to use would be only 8 cm³. This small volume would not mix so readily or effectively with the oxide catalyst.

2 (a) (i) Action of $NH_3(aq)$ and $NH_4Cl(aq)$ as a buffer.
In the solution,
$NH_3(aq)$ forms an equilibrium,

$$NH_3 + H_2O \rightleftharpoons NH_4^+ + OH^- \qquad (1) \quad \text{(equation 1)}$$

and $NH_4Cl(aq)$ is fully dissociated into ions,

$$NH_4Cl(aq) \rightarrow NH_4^+(aq) + Cl^-(aq) \qquad (1) \quad \text{(equation 2)}$$

If a small amount of acid ($H^+(aq)$) is added, the extra H^+ ions react with OH^- ions to form H_2O. (1)
The equilibrium in equation 1, then moves to the right to restore the $[OH^-(aq)]$ and the pH does not change very much. (1)

If a small amount alkali (OH^-) is added, the extra OH^- ions react with NH_4^+ (1) and the equilibrium in equation 1 moves to the left. Extra OH^- is therefore removed and the pH does not change very much. (1)

(ii) edta and Ca^{2+} ions react in a ratio of 1 : 1.

Amount of edta reacting $= \dfrac{31.2}{1000} \times 0.01 = 3.12 \times 10^{-4}$ moles (1)

$\Rightarrow 3.12 \times 10^{-4}$ moles Ca^{2+} in 50 cm³ tap water

$\therefore$ calcium ion concentration $= 3.12 \times 10^{-4} \times \dfrac{1000}{50}$ (1)

$= 6.24 \times 10^{-3}$ mol dm⁻³ (1)

(b) (i) $Ca^{2+}(aq) + C_2O_4{}^{2-}(aq) \rightarrow CaC_2O_4(s)$ (1)

No. of moles of $Ca^{2+}(aq)$ taken $= \dfrac{25}{1000} \times 0.05 = 1.25 \times 10^{-3}$

$\Rightarrow$ moles of CaC_2O_4 ppte $= 1.25 \times 10^{-3}$ (1)

$\therefore$ mass of CaC_2O_4 ppte $= 1.25 \times 10^{-3} \times$ molar mass of CaC_2O_4
$= 1.25 \times 10^{-3} \times 128$
$= 0.160$ g (1)

> **Examiner's tip**
> In calculations related to reactions, always write the equation for the reaction if it has not been provided in the question.

(ii) $\left.\begin{array}{l} MnO_4{}^- + 8H^+ + 5e^- \rightarrow Mn^{2+} + 4H_2O \times 2 \\ C_2O_4{}^{2-} \rightarrow 2CO_2 + 2e^- \qquad\qquad\qquad \times 5 \end{array}\right\}$ (1)

$\Rightarrow$ 2 mol $MnO_4{}^-$ react with 5 mol $C_2O_4{}^{2-}$

or $2MnO_4{}^- + 16H^+ + 5C_2O_4{}^{2-} \rightarrow 2Mn^{2+} + 8H_2O + 10CO_2$ (1)

moles of $C_2O_4{}^{2-}$ in 250 cm³ solution $= 1.25 \times 10^{-3}$

$\Rightarrow$ moles of $C_2O_4{}^{2-}$ reacting $= 1.25 \times 10^{-3} \times \dfrac{25}{250}$

$= 1.25 \times 10^{-4}$ (1)

$\therefore$ Moles of $MnO_4{}^-$ reacting $= 1.25 \times 10^{-4} \times \dfrac{2}{5}$ (1)

If the volume of 0.002 mol dm⁻³ $KMnO_4$ reacting is V cm³,

$\dfrac{V}{1000} \times 0.002 = 1.25 \times 10^{-4} \times \dfrac{2}{5}$ (1)

$V = 25$ cm³ (1)

> **Examiner's tip**
> In redox reactions, it is often easier to write the two half-equations (one involving oxidation, the other reduction) and then balance the half-equations so that equal numbers of electrons are taken by the oxidant and given up by the reductant.

3 (a) the ester group/—C—O— (1)
$\qquad\qquad\qquad\;\;\; \overset{\displaystyle\|}{O}$

(b) (i) Moles of methyl benzoate $= \dfrac{2.70}{\text{molar mass}}$

$= \dfrac{2.70}{136}$

$= 0.0199$ (1)

(ii) Concentration of NaOH is 4.0 g in 50 cm^3 solution.

$$\Rightarrow 4.0 \times \frac{1000}{50} = 80 \text{ g dm}^{-3} (1) = \frac{80}{40} \text{ mol dm}^{-3} = 2.0 \text{ mol dm}^{-3}$$
$$(1)$$

(iii) From the equation, 1 mole of methyl benzoate gives 1 mole of benzoic acid   (1)

∴ theoretical yield of benzoic acid = 0.0199 moles   (1)
= 0.0199 × 122
= 2.43 g   (1)

$$\Rightarrow \text{ percentage yield } = \frac{1.50}{2.43} \times 100$$

$$= 61.7\% \ (1)$$

> **Examiner's tip**
> *The mass of methyl benzoate was given to 3 significant figures. Consequently, the number of moles of methyl benzoate in (b)(i) has been calculated to 3 significant figures.*
>
> *On the other hand, the mass of sodium hydroxide is only given to 2 significant figures, so its concentration in (b)(ii) is calculated to just 2 significant figures. The sodium hydroxide is, however, present in large excess, so it will not influence the yield of benzoic acid, which can be given to 3 significant figures in (b)(iii).*

(iv) Some of the methyl benzoate may not have reacted (hydrolysed) fully.   (1)

Some benzoic acid is lost (dissolved) in the aqueous solution when the crystals are filtered off.   (1)

Some benzoic acid is also lost, remaining dissolved in the water during recrystallisation.   (1)   (any 2 points for 2 marks)

(c) (i) After refluxing, the products will be methanol and sodium benzoate. The sodium benzoate is soluble in the aqueous mixture.   ($\frac{1}{2}$)
On adding dilute sulphuric acid, H$^+$ ions react with benzoate ions and form insoluble benzoic acid.   ($\frac{1}{2}$)

$$C_6H_5COO^-(aq) + H^+(aq) \rightarrow C_6H_5COOH(s)$$

(ii) To purify them ($\frac{1}{2}$) and ensure they are not contaminated with methyl benzoate, sodium benzoate, sulphuric acid or sodium sulphate.   ($\frac{1}{2}$)

(d) (i) Methyl benzoate: ester $-\overset{\displaystyle \|}{\underset{\displaystyle O}{C}}-$ bond 1750–1735 cm^{-1}

or ester C–O bond 1310–1250 cm^{-1} } (1)

Benzoic acid: acid C–O bond 1700–1680 cm^{-1} (1)
acid O–H bond 3300–2500 cm^{-1} (1)
Methanol: alcohol O–H bond 3750–3200 cm^{-1} (1)

(ii) Check that it has no absorptions corresponding to the aryl C–H bond at 3030 cm^{-1} (1), acid C–O bond at 1700–1680 cm^{-1} or acid O–H bond at 3300–2500 cm^{-1}.

> *Examiner's tip*
> *Make sure that you are familiar with the data book or data sheets that may be available to you in some of your unit tests. This is particularly important for looking up IR absorption wavenumbers, as in this question, standard electrode (reduction) potentials ($E^{\ominus}$), standard enthalpy changes of formation ($\Delta H_f^{\ominus}$) and standard entropies ($S^{\ominus}$).*

Paper 5 (Unit Test 5) Mark Scheme and Examiner's Tips

1 Boiling points

NaCl consists of <u>Na$^+$ and Cl$^-$ ions</u> (1) held together by the electrical attraction of <u>strong ionic bonds</u> (1) in a <u>giant structure</u> (1). Hence the boiling point is high.
(>1500 K) (1)
The structures of CH_3COOH, CH_3CH_2OH and $AlCl_3$ consist of discrete <u>simple molecules</u>. (1) There are strong covalent bonds holding the atoms within the separate molecules but only <u>relatively weak intermolecular forces between the separate molecules</u>. (1) Hence, <u>their boiling points are relatively low</u> (c 400 K) (1) compared to that of NaCl.

Electrical conductivity of aqueous solutions

In aqueous solution, <u>NaCl dissociates to form freely moving Na$^+$ and Cl$^-$ ions</u>. (1) When electrodes are placed in a solution of NaCl(aq), <u>these ions can move</u> to the oppositely charged electrodes and <u>give good conductivity</u> (1) to an electric current.
<u>$AlCl_3$ reacts with water</u> (1) to form a solution containing $[Al(H_2O)_6]^{3+}$(aq), Cl$^-$(aq), $[Al(H_2O)_5OH]^{2+}$(aq) and H$^+$(aq).
Like NaCl(aq), <u>these ions provide good electrical conductivity</u>. (1)
CH_3COOH (ethanoic acid) is <u>only partially dissociated (ionised)</u> (1) in aqueous solution. The undissociated CH_3COOH molecule is in equilibrium with smaller amounts of CH_3COO^- and H_3O^+ ions.

$$CH_3COOH(aq) \rightleftharpoons CH_3COO^-(aq) + H^+(aq) \quad (1)$$

The small amounts of aqueous ions result in slight conductivity of CH_3COOH(aq).
CH_3CH_2OH mixes completely with water, but <u>does not form ions</u>. (1) Its conductivity is poor – no better than that of pure water.

Concentration of H$^+$ ions in aqueous solutions

NaCl dissolves in water to form Na$^+$(aq) and Cl$^-$(aq) ions. <u>It does not disturb the equilibrium of the water.</u> (1)

$$H_2O(aq) \rightleftharpoons H^+(aq) + OH^-(aq)$$
$$\text{in which } [H^+(aq)] = 1.0 \times 10^{-7} \text{ mol dm}^{-3} \text{ at } 25°C \Bigg\} \quad (1)$$

$AlCl_3$ dissolves in water forming $[Al(H_2O)_6]^{3+}$(aq) and Cl^-(aq) ions. <u>The $[Al(H_2O)_6]^{3+}$(aq) then reacts further to form H^+(aq) ions.</u>   (1)

$$[Al(H_2O)_6]^{3+}(aq) \rightleftharpoons [Al(H_2O)_5OH]^{2+}(aq) + H^+(aq) \quad (1)$$

$$\text{or } [Al(H_2O)_6]^{3+}_{(aq)} + H_2O_{(1)} \rightleftharpoons [Al(H_2O)_5OH]^{2+}_{(aq)} + H_3O^+_{(aq)}$$

<u>This reaction occurs due to the high positive charge density on the central Al^{3+} ion</u> (1) causing the ejection of an H^+ ion. The additional H^+ ions in the solution raise the $[H^+]$ to 3.0×10^{-1} mol dm^{-3}.

As already stated, CH_3COOH is partially ionised in aqueous solution. A small fraction of the 0.1 mol dm^{-3} solution ionises and this results in a concentration of H^+(aq) of 1.3×10^{-3} mol dm^{-3}. (Award marks here if not obtained earlier.)

CH_3CH_2OH dissolves in water, but a reaction does not occur. <u>The $[H^+]$ in aqueous CH_3CH_2OH is therefore the same as that in pure water.</u>   (1)

(max 14 marks)

> **Examiner's tip**
> The logical approach to this question is to deal with the three properties in turn, comparing the four substances. Remember to keep to the question. Don't waffle, keep your answer relevant and don't take more than about 15 minutes on the question. You will see from the mark scheme that there is a total of 20 possible marks, so 15 minutes to obtain a maximum of 14 is not unreasonable.

2 (a) (i) Since $\Delta H^{\ominus}_f$ (C_2H_4(g)) is positive and $\Delta H^{\ominus}_f$ (C_2H_6(g)) is negative, $\Delta H_{reaction}$ is positive (endothermic)   (1)
∴ applying Le Chatelier's principle, an <u>increase in temperature will move the equilibrium to the right products) and this will increase the value of K_p.</u>   (1)

(ii) Applying Le Chatelier's principle; increase in pressure will move the equilibrium to the left, but this will change the actual pressures of the substances themselves and the <u>value of K_p will not change.</u>   (1)

(b)

Initial amounts /moles	C_2H_6(g)	$\rightleftharpoons$	C_2H_4(g)	+	H_2(g)	
	1.0		0		0	(Total no.
Final amounts /moles	0.64		0.36		0.36	of moles at equilibrium

(1)   = 1.36) (1)

Final pressure /kPa $\dfrac{0.64}{1.36} \times 180$ $\dfrac{0.36}{1.36} \times 180$ $\dfrac{0.36}{1.36} \times 180$

= 84.70   (1)   = 47.65   = 47.65

$K_p = \dfrac{(P_{C2H4})_{eqm} \times (P_{H2})_{eqm}}{(P_{C2H6})_{eqm}}$ (1)   (1)

$= \dfrac{(47.65)^2}{84.70}$

= 26.8 (1) kPa   (1)

3 <u>What is meant by a redox reaction?</u>

Redox involves reduction and oxidation.
Initially, oxidation was defined as addition of oxygen and reduction was defined as removal of oxygen.   (1)
∴ in the reaction:

$$Fe_2O_3 + 2Al \rightarrow 2Fe + Al_2O_3$$

Fe_2O_3 is reduced and Al is oxidised.   (1)
But when Al changes to Al_2O_3, the <u>Al atoms become Al^{3+} ions</u> and, at the same time, <u>Fe^{3+} ions in Fe_2O_3 become Fe atoms</u>.   (1)
(or another reaction to illustrate redox in terms of electron loss and gain)

So, <u>$Al \rightarrow Al^{3+} + 3e^-$ is oxidation</u>   (1)
i.e. <u>oxidation is the loss of electrons</u>   (1)
and <u>$Fe^{3+} + 3e^- \rightarrow Fe$ is reduction</u>   (1)
i.e. <u>reduction is the gain of electrons</u>   (1)
(or deduced from/related to another reaction)

Defining redox in terms of electron loss or gain was very helpful in inorganic chemistry, but less so in organic chemistry where most substances are not ionic and reactions cannot usually be understood in terms of electron transfer.

The concept of oxidation number was therefore developed and in terms of redox:

- <u>reduction involves a decrease in oxidation number</u>   (1) and
- <u>oxidation involves an increase in oxidation number</u>.   (1)

$$\text{Oxidation numbers} \quad \overset{-4}{C} \overset{+1}{H_4} \; + \; \overset{0}{Cl_2} \; \rightarrow \; \overset{-2}{C} \overset{+1}{H_3} \overset{-1}{Cl} \; + \; \overset{+1}{H} \overset{-1}{Cl} \quad (1)$$

So, in this reaction above, carbon is oxidised from oxidation number -4 to -2;   (1)
chlorine is reduced from oxidation number 0 to -1.   (1)
(or another reaction to illustrate redox in terms of oxidation numbers)

Another two reactions to illustrate redox in terms of electron transfer or change in oxidation numbers (each reaction carries 3 marks, (1) for the equation, (1) for pin-pointing oxidation, (1) for pin-pointing reduction. (At least 2 of the illustrative reactions from inorganic chemistry and at least 2 illustrative reactions from organic chemistry.)   (max 14 marks)

> **Examiner's tip**
> Notice in the mark scheme how the question has been answered by discussing the development of the concept of redox from addition/removal of oxygen through loss/gain of electrons to increase/decrease in oxidation numbers.
>
> It is always helpful and effective to give some such structure to nebulous questions like this one.

4 **(a)** $CH_4(g) + Cl_2(g) \rightarrow CH_3Cl(g) + HCl(g)$ (1)
(state symbols not essential)

Mechanism

Initiation: $Cl_2 \xrightarrow{\text{UV}} Cl\cdot + Cl\cdot$ (1) (chlorine free-radicals)

Propagation: $Cl\cdot + CH_4 \rightarrow CH_3\cdot + HCl$ (1)
          $CH_3\cdot + Cl_2 \rightarrow CH_3Cl + Cl\cdot$ (1)

Termination: $Cl\cdot + Cl\cdot \rightarrow Cl_2$ (this reaction is unlikely as methane is in excess)
          $Cl\cdot + CH_3\cdot \rightarrow CH_3Cl$ (1)
     or $CH_3\cdot + CH_3\cdot \rightarrow CH_3CH_3$

> **Examiner's tip**
> In ultraviolet-induced reactions, the initiation process usually involves the formation of halogen free-radicals, e.g. Cl·.
>
> Propagation then commences with Cl· attacking CH_4 to produced HCl and $CH_3\cdot$ radicals.
>
> Termination can be brought about by the combination of any two free-radicals.

(b) Direct hydration

$C_2H_4(g) + H_2O(g) \rightarrow C_2H_5OH(g)$
H_3PO_4 (phosphoric acid) catalyst   (1)
High temperature c.200°C (500 K)   (1)
High pressure 70 atm   (1)
(If you have simply said 'high temperature' and 'high pressure' without specifying values, this would gain only 1 mark in total)

Fermentation

$C_6H_{12}O_6(aq) \rightarrow 2C_2H_5OH(aq) + 2CO_2(g)$
In the presence of yeast   (1)
Atmospheric pressure   (1)
Temperature c.35°C (310 K)   (1)
Anaerobic conditions (no air)   (1)   (max 3 marks)

Direct hydration

Advantages – any two of: cheaper (continuous and fast process)   (1)

fast process, catalysed at high temperature and pressure   (1)

a continuous process   (1)

product can be very pure   (1)   (max 2)

Disadvantage –   uses up a valuable, finite resource (crude oil from which C_2H_4 is obtained)   (1)

Examiner's tip
Notice that there are no marks in part (b) for writing the equations for direct hydration and fermentation. This is because they are not asked for in the question. This is a very good reminder that even in essay questions, you should initially do only what you are asked to do in the question. If you are left with time to spare, then go back and decide whether there is more that you could add to your answers.